"This book makes a timely contribution towards underst[...] and practicality of climate neutral and resilient farming s, smallholder farms, and the potential for climate mitigation and the Nationally Determined Contributions (NDCs). It adds new knowledge and is useful for a wide audience in this field."

Dr. V. Geethalakshmi, *Ph.D., FAAM,*
Vice-Chancellor, Tamil Nadu Agricultural University,
Coimbatore, India

"IPCC's 6th Assessment Report of Working Group I stated in 2021 that 'It is unequivocal that human influence has warmed the atmosphere, ocean and land'. The agricultural sector should take immediate measures to mitigate and adapt to the climate change due to global warming. For irrigation engineers, it is very significant to have a firm grasp of the topic Climate Neutral and Resilience Farming Systems (CNRFSs), which is the focus of this book."

Tsugihiro Watanabe, *Vice President of the International*
Commission on Irrigation and Drainage (ICID), Professor
Emeritus of Kyoto University, Japan

"This book is very timely and a valuable knowledge contribution on managing green-house-gas emissions in the agriculture sector. Each chapter underscores low carbon emission agriculture in achieving the 1.5°C target of the Paris Agreement. An insightful publication!"

Paxina Chileshe-Toe, *Regional Climate and Environment*
Specialist, Environment, Climate, Gender and Social
Inclusion Division, Strategy and Knowledge Department,
IFAD, Nairobi, Kenya

"We need to deploy innovative solutions at scale to tackle climate change. This book is timely in helping to promote agri-food systems as an important part of the solution to the climate crisis, especially in the lead up to COP27 in Egypt and beyond."

Zitouni Ould-Dada, *Deputy Director, Office of Climate*
Change, Biodiversity and Environment (OCB), Food and
Agriculture Organization of the United Nations (FAO), Via
delle Terme di Caracalla, Rome, Italy

Climate Neutral and Resilient Farming Systems

This book presents evidence-based research on climate-neutral and resilient farming systems and further provides innovative and practical solutions for reducing greenhouse gas emissions and mitigating the impact of climate change.

Intensive farming systems are a significant source of greenhouse gas emissions, thereby contributing to global warming and the acceleration of climate change. As paddy rice farming is one of the largest contributors, and environmentally damaging farming systems, it will be a particular focus of this book. The mitigation of greenhouse gas emissions needs to be urgently addressed to achieve the 2°C target adopted by COP21 and the 2015 Paris Agreement, but this is not possible if local and national level innovations are not accompanied by international level cooperation, mutual learning and sharing of knowledge and technologies. This book, therefore, brings together international collaborative research experiences on climate-neutral and resilient farming systems compiled by leading scientists and experts from Europe, Asia and Africa. The chapters present evidence-based research and innovative solutions that can be applied or upscaled in different farming systems and regions across the world. Chapters also present models and technologies that can be used for practical implementation at the systemic level and advance the state of the art knowledge on carbon-neutral farming. Combining theory and practice, this interdisciplinary book provides guidance which can inform and increase cooperation between researchers from various countries on climate-neutral and resilient farming systems. Most importantly, the volume provides recommendations which can be put into practice by those working in the agricultural industry, especially in developing countries, where they are attempting to promote climate-neutral and resilient farming systems.

The book will be of great interest to students and academics of sustainable agriculture, food security, climate mitigation and sustainable development, in addition to policymakers and practitioners working in these areas.

Udaya Sekhar Nagothu is Research Professor and Director at the Centre for International Development, NIBIO (Norsk Institutt for Biookonomi/Norwegian Institute of Bioeconomy Research), Norway. He is the editor of *The Bioeconomy Approach* (2020), *Agricultural Development and Sustainable Intensification* (2018), *Climate Change and Agricultural Development* (2016) and *Food Security and Development* (2015).

Earthscan Food and Agriculture

Climate Neutral and Resilient Farming Systems

Practical Solutions for Climate Mitigation and Adaptation

Edited by
Udaya Sekhar Nagothu

Routledge
Taylor & Francis Group
LONDON AND NEW YORK

earthscan
from Routledge

First published 2023
by Routledge
4 Park Square, Milton Park, Abingdon, Oxon OX14 4RN

and by Routledge
605 Third Avenue, New York, NY 10158

Routledge is an imprint of the Taylor & Francis Group, an informa business

British Library Cataloguing-in-Publication Data
A catalogue record for this book is available from the British Library

Library of Congress Cataloging-in-Publication Data
Names: Nagothu, Udaya Sekhar, editor.
Title: Climate neutral and resilient farming systems : practical solutions for climate mitigation and adaption / edited by Udaya Sekhar Nagothu.
Description: New York, NY : Routledge, 2023. | Includes bibliographical references and index.
Identifiers: LCCN 2022021933 (print) | LCCN 2022021934 (ebook)
Subjects: LCSH: Sustainable agriculture. | Crops and climate. | Agriculture—Environmental aspects.
Classification: LCC S600.5 .C556 2023 (print) |
LCC S600.5 (ebook) | DDC 630—dc23/eng/20220722
LC record available at https://lccn.loc.gov/2022021933
LC ebook record available at https://lccn.loc.gov/2022021934

ISBN: 978-1-032-22579-1 (hbk)
ISBN: 978-1-032-22584-5 (pbk)
ISBN: 978-1-003-27317-2 (ebk)

DOI: 10.4324/9781003273172

Typeset in Goudy
by codeMantra

Contents

Preface

We are now facing an increase in the frequency of extreme weather events due to climate change, which is exacerbating temperature extremes and impacting soil, water and growth conditions of crops. The agriculture sector is both a victim and cause of climate change. To address the climate crisis, we need a transformative change in the way we farm in the future and move from the intensive farming systems towards carbon-neutral farming. The change implies a drastic reduction in the use of external chemical inputs and adopting agroecological-based practices wherever possible. Any climate mitigation efforts to reduce greenhouse gas emissions in agriculture and food systems must benefit other relevant sectors and provide co-benefits to adaptation and resilience.

There is no "one-size-fits-all" solution to address climate crisis. A mosaic of adaptation and mitigation options that suit different situations considering the environmental, social and economic contexts and vulnerabilities must be developed. The package of measures must include nature-based, cultural, physical and biological solutions suitable for the agroecosystems. Further, the efforts need supportive policies, collective stakeholder action, knowledge sharing and adequate investments to promote systematic implementation. Though the limited funding opportunities in developing countries will force governments to follow the economic agenda rather than invest in climate action, there is still hope. One way to address this challenge is by ensuring that development work is "climate proofed" and climate action to be development oriented. In this way, governments can justify their investments to combat climate change.

The various chapters in the book were drafted by 33 experienced researchers and consultants from several disciplines representing more than 20 agencies worldwide, bringing together diverse field experiences. Several of the book chapters focus on rice, the major cereal providing food security to millions of people worldwide. Paddy rice is also one of the major sources of methane emissions and facing several challenges due to high input prices, increased incidence of pests, low market prices and labour shortages. The book emphasizes on the relevance and use of agroecological-based soil, water and crop management practices that have the potential for increasing productivity whilst reducing greenhouse gases and addressing relevant SDGs (especially SDGs 2, 13 and 15). Addressing the climate-related challenges will not be easy unless the farmers are motivated, in-centivized and willing to adapt to the change. We must be optimistic, as it is necessary to make farming systems resilient, and at the same time mitigate future climate risks. The open access book will be useful to a wide range of audience including scientific community, development agencies and policymakers.

Figures

Tables

Boxes

Contributors

Bhubananda Adhikari is Research Scholar at Odisha University of Agriculture & Technology, Bhubaneswar, India.

Laura Bardi is Director of Research at the Research Centre for Engineering and Agro-Food Processing, Council for Agricultural Research and Economics (CREA), Italy.

Chiara Bertora is agroecologist and former researcher in the Department of Agricultural, Forest and Food Sciences (DISAFA), University of Turin, Italy.

Patrizia Borsotto is Researcher at the Research Centre for Agricultural Policies and Bioeconomy of the Council for Agricultural Research and Economics (CREA), Italy.

Andrew Borrell is Professor at Queensland Alliance for Agriculture and Food Innovation (QAAFI), Hermitage Research Facility, University of Queensland, Warwick, Australia.

Giacomo Branca is Associate Professor in the Department of Economics, Engineering, Society and Business, Tuscia University, Italy.

Luca Cacchiarelli is Researcher in the Department of Economics, Engineering, Society and Business, Tuscia University, Italy.

Roberto Cagliero is Researcher at the Research Centre for Agricultural Policies and Bioeconomy of the Council for Agricultural Research and Economics (CREA), Italy.

Alamu Oladeji Emmanuel is Food Scientist working at the International Institute for Tropical Agriculture (IITA), Zambia office.

Omedé Gabriele is Research Fellow in the Department of Agricultural, Forest and Food Sciences (DISAFA), University of Turin, Italy.

Jan Willem Ketelaar is Chief Technical Advisor of the FAO Regional Integrated Pest Management/Pesticide Risk Reduction Programme at FAO's Regional Office for Asia and Pacific, Bangkok.

Anjani Kumar is Senior Scientist in the Division of Crop Production, National Rice Research Institute (ICAR), Cuttack, India.

Maite Martínez-Eixarch is Researcher at the Institute of Agrifood Research and Technology (IRTA), Continental and Marine Waters, Spain.

Abha Mishra is Director of ACISAI Center Asian Institute of Technology, Khlong Luang, Thailand.

Soken Matsuda is Researcher at National Agriculture and Food Research Organization (NARO), Japan.

Sangita Mohanty is Senior Scientist in the Division of Crop Production, National Rice Research Institute (ICAR), Cuttack, India.

Shyamaranjan Das Mohapatra is Principal Scientist in the Division of Crop Protection, National Rice Research Institute (ICAR), Cuttack, India.

Kiran Mohapatra is Research Scholar in the Division of Crop Production, National Rice Research Institute (ICAR), Cuttack, India.

Minati Mohapatra is Assistant Research Engineer at Odisha University of Agriculture & Technology, Bhubaneswar, India.

Munmun Mohapatra is Research Scholar at Odisha University of Agriculture & Technology, Bhubaneswar, India.

Stefano Monaco is Researcher at the Research Centre for Engineering and Agro-Food Processing, Council for Agricultural Research and Economics (CREA), Italy.

Udaya Sekhar Nagothu is Research Professor and Director of Centre for International Development, Norwegian Institute of Bioeconomy Research, Ås, Norway.

Kimihito Nakamura is Professor of Hydrological Environment Engineering in the Graduate School of Agriculture, Kyoto University, Japan.

Amaresh Kumar Nayak is Principal Scientist and Head Division of Crop Production at National Rice Research Institute, Cuttack, India.

Joel Ngumayo is Senior M&E Specialist/Researcher at Community Markets for Conservation (COMACO), Lusaka, Zambia.

Joyce Bakuwa Njoloma is Associate Scientist – Systems at World Agroforestry, Malawi and Associate Professor at the Lilongwe University of Agriculture and Natural Resources (LUANAR), Lilongwe, Malawi.

Chiara Perelli is Junior Researcher in the Department of Economics, Engineering, Society and Business, Tuscia University, Italy.

Subhendu Sekhar Pradhan is Computer Assistant in the Division of Crop Protection, National Rice Research Institute (ICAR), Cuttack, India.

Le Xuan Quang is Associate Professor and Vice Director at the Institute for Water and Environment, Vietnam Academy for Water Resources, Vietnam.

Radhakrushna Senapati is Senior Research Fellow in the Division of Crop Protection, National Rice Research Institute (ICAR), Cuttack, India.

Mehreteab Tesfai is Senior Researcher at the Norwegian Institute of Bioeconomy Research, Ås, Norway.

Rahul Tripathi is Senior Scientist in the Division of Crop Production, National Rice Research Institute (ICAR), Cuttack, India.

Max Whitten is Adjunct Professor in the School of Biological Sciences, The University of Queensland, Brisbane, Australia.

Najam Waris Zaidi is Senior Associate Scientist II at International Rice Research Institute, New Delhi, India.

1 Climate change impacts on agriculture

Challenges and options to reduce emissions through climate-neutral and resilient farming systems

Udaya Sekhar Nagothu, Andrew Borrell, and Mehreteab Tesfai

Introduction

A global *climate crisis* is drawing the attention of activists, politicians, scientists, and the general public at large, not only due to the increasing rate of extreme climate events across the world and the severity of the destruction caused by these events to communities and ecosystems but also due to their continuous coverage in the media (WMO, 2021). At the same time, climate change debate is shaping the political landscape in several countries, with some countries seriously concerned and pressing for immediate action, while others do not see it as an immediate threat, even in the developed world. Lack of adequate information, evidence-based data, and uncertainty in forecasts are helping sceptics and politicians in both developed and developing countries to argue that climate change is not an immediate threat to global society. Such ignorance leads to short-sighted policy decisions and lack of needed transformative action and support for investments to combat climate crises. Since 1990, six assessment reports by the Intergovernmental Panel on Climate Change (IPCC) were prepared, and recommendations were made for cutting down greenhouse gas (GHG) emissions (IPCC, 2021). Unfortunately, some world leaders do not recognize the seriousness of the threats and fail to stand by the commitments made to reduce emissions. As long as these commitments are not put into action, it will not be possible to limit global temperature rise to 1.5°C by the end of the century.

The IPCC on the Sixth Assessment Report (AR6) states that "it is unequivocal that human influence has warmed the atmosphere, ocean and land" and that "widespread and rapid changes in the atmosphere, ocean, cryosphere and biosphere have occurred" (IPCC, 2021a). According to the Report, the world has rapidly warmed by 1.1°C which is higher than pre-industrial levels, and is now moving towards 1.5°C – a critical threshold level that world leaders agreed to maintain and take measures to prevent warming above that level (IPCC, 2021b). The complex shifts observed in recent years affecting our planet's weather and climate systems are contributing to the melting of glaciers, sea-level rise, and

DOI: 10.4324/9781003273172-1

increase in temperature. The atmospheric carbon dioxide levels reached a record high in 2020, unprecedented in human history as the world was also grappling with one of the worst pandemics (NORR, 2021). The year 2020 was also one of the hottest years recorded globally, and the hottest ever in Europe that has led to serious forest fires and floods (WMO, 2021). The wild fires in California and Australia, the destructive floods in Germany, and the heat waves in Canada during the summer of 2021 all indicate that the climate crisis is impacting seriously and can no longer be ignored (GDACS, 2021). The scale of destruction not only to property and infrastructure but also to human life and ecosystems cannot continue to be tolerated, especially in regions and populations that are highly vulnerable. Limiting global warming is only possible by taking drastic measures to cut GHG emissions, while also removing carbon dioxide from the atmosphere through large-scale carbon sequestration measures (IPCC, 2021b). The economic instability caused by COVID-19, with the focus by governments on funding health initiatives combined with priorities to ensure jobs and economic growth, will nevertheless pose a big challenge to combat the climate crisis.

This chapter provides an introduction to the climate crisis, followed by a brief overview of the sources and extent of GHG emissions from various sectors in general, and the agriculture sector in particular, and the challenges to reduce emissions from the latter. The chapter then discusses the potential solutions for reducing GHG emissions from the agriculture sector, including technological, investment, and policy support required. A separate section is dedicated to introduce the climate-neutral and resilient farming systems (CNRFS) concept and the various steps necessary for assessing and developing suitable CNRFS. Towards the end, the chapter provides an outline on the various chapters of the book.

Global warming and contribution from various sources

The sectors that contribute most to GHGs and global warming are energy, industry, agriculture, transport, and associated land-use changes. Per capita or per person emissions of GHGs are the highest in the USA followed by Russia, Japan, China, and the European Union (EU), ranked in the order given (C2ES, 2021). In terms of contribution to total emissions, China ranks the highest followed by the USA, the European Union, India, and Russia. Overall, carbon dioxide accounts for 76% of total GHGs, while methane emissions, primarily from the agriculture sector, contribute 16% of the GHGs (C2ES, 2021). The third largest contributor is nitrous oxide, contributing 6% of the emissions mainly from industry and agriculture, and the remaining 2% by various carbon and nitrogenous compounds (e.g. CO, NO). Reducing emissions from various sectors has to be a collective effort since the consequences of climate change are global.

Countries such as India argue that since its per capita emissions are much lower compared to other nations, its responsibility to invest and reduce GHGs should accordingly be less, a position with which other industrial nations do not agree (UNEP, 2020). Lack of consensus to reduce emissions does not help to move forward, as evident during various meetings since the Kyoto Protocol was established in 1997. The Kyoto Protocol set targets for 37 industrial nations to cut down emissions, but only bound to developed countries under the principle of "common but differentiated responsibility and respective capabilities", as these countries were also responsible for the current high levels of GHG emissions (UNFCCC, 2021). At the Doha Amendment in 2012, the Protocol was adopted for the second commitment period ending in 2020, with the aim of reducing emissions by 18%. However, the results so far are not encouraging, with only a few scattered transformative actions undertaken. The outcomes and agreements reached at the COP26 summit in Glasgow (Scotland) are crucial for the global community to hit net-zero emissions (WSJ, 2021). The years and months that follow will tell us the extent to which the COP26 commitments will be acted on.

The agriculture sector as a contributor to GHG emissions

Agriculture accounts for approximately 20% of global GHG emissions when viewed over a 20-year time frame, while forestry and land-use change account for about 7%. Hence, the combined contributions from agriculture, forestry, and changes to land use account for more than one-quarter of the world's GHG emissions. Without targeted action, these emissions are likely to increase as the population increases and the demand for food and raw materials continues to grow.

Agriculture contributes substantially to climate change by directly emitting non-carbon dioxide (non-CO_2) gases, including methane (CH_4) and nitrous oxide (N_2O), from crop and livestock production, and by affecting net CO_2 emissions from agricultural soils, forestry, and other land use (OECD, 2021). After livestock, rice production is the second largest contributor of agricultural CH_4 emissions, with the remaining emissions from the burning of savanna and the use of crop residues for agricultural purposes. Furthermore, agriculture accounts for 80% of total N_2O emissions, mainly from the application of fertilizers, including both synthetic and organic nitrogen (Reay et al., 2012). These non-CO_2 gases are significantly more powerful than carbon dioxide in driving warming over a 20-year span (Myhre et al., 2013). For example, over a span of 20 years, methane is 84 times more powerful than CO_2 in forcing temperature increases, and nitrous oxide is 298 times more powerful. However, CH_4 has a much shorter lifetime in the atmosphere, lasting only 12 years, meaning that reducing CH_4 emissions should help to limit temperature increases in the short term, which is necessary.

Estimates of global CH_4 emissions from paddy fields alone range from 31 to 112 Tg/year, accounting for up to 19% of total emissions, while 11% of global agricultural N_2O emissions come from rice fields (Win et al., 2020). In Asia, paddy rice systems alone (besides livestock and energy use) are the second largest emitter of GHG emissions. At the same time, GHG emissions from rice cultivation are increasing in other rice growing regions of the world. Rice is the staple food for more than half of the world's population and the dominant crop in South and Southeast Asia (FAO, 2013). More than 3.5 billion people depend on rice for nearly 20% of their daily calories (FAO, 2018). Therefore, rice production has to be made sustainable through use of viable technology options to reduce GHG emissions, especially methane. Reducing emissions from the agriculture sector – largely methane and nitrous oxide – can play a significant role in climate change mitigation (Lynch and Garnett, 2021).

The current farming systems worldwide are intensive and unsustainable, as they mostly rely on high external inputs to increase crop yields, e.g. water, chemical fertilizers, and pesticides, and contribute to serious degradation of soils, pollution of water and air, and loss of biodiversity and ecosystem services (Foley et al., 2011). In general, intensive agricultural land-use systems are a major source of GHG emissions, thereby contributing to the acceleration of climate change (FAO, 2020). This, in turn, results in frequent extreme weather events, i.e. droughts, floods, and heat stress (Mwangi et al., 2020). Climate warming is substantially increasing crop losses due to the increased spread of insect pests, with strongest yield reductions in temperate and subtropical climatic zones for major staple grains including rice, maize, and wheat (Boetzl et al., 2020; Deutsch et al., 2018). These studies highlight that the agricultural sector is both a contributor to and casualty of accelerated climate change, and that there is an urgency to promote new systemic solutions to promote sustainable CNRFS.

The challenges in reducing GHG emissions

The 2018 report by IPCC makes it clear that if the impact of climate change is to be limited to 1.5°C, a "rapid and far-reaching" transition will be required (IPCC, 2018). Achieving this goal would require keeping within the cumulative carbon budget of 570 gigatonnes of carbon dioxide ($GtCO_2$), attaining net-zero carbon dioxide emissions globally around 2050, including significantly reducing the emissions of other gases such as methane and nitrous oxide.

The focus on the agriculture sector to reduce GHGs has increased since the Paris Agreement in 2015, where 196 countries signed a legally binding treaty on climate change to limit global warming to well below 2°C, and preferably below 1.5°C (UNFCCC, 2015). It was also agreed in Paris that countries prepare their plans for nationally determined contributions (NDCs) by 2020, together with the long-term low GHG emission development strategies (LT-LEDS) embedded into the NDCs. However, practical implementation of NDCs is still at a nascent stage in most cases. So far, the attempts to reduce emissions are not encouraging in agriculture, livestock, and land use sectors. It can be challenging for sectors such as agriculture to monitor and measure the emission reductions due to the diffused and unorganized nature of the small-scale farms. In light of the impact of the COVID-19 pandemic and shifting geo-political power alignments, the question of who should support and invest in climate mitigation measures remains deeply contested. While debates at the global climate negotiation meetings are often focused on the extreme and contentious views and options, there should be a spectrum of nuanced views that will be useful to look at and move beyond the contentious positions. A transparent negotiating environment will help developing countries to engage actively in reducing emissions. This should be particularly useful for sectors such as agriculture that are highly relevant in these countries.

Reducing emissions within the agriculture sector will be challenging, mainly due to lack of investments and capacity, small-scale nature of farming, lack of political will, and incentives for adopting climate smart solutions (FAO, 2015). Evidence-based and proven practical solutions to reduce emissions in the agriculture sector are still relatively rare and some are in the piloting stage. The type of solutions that we develop will have implications for overall food systems and for global food and nutrition security at large. We need to understand that emissions from agriculture are different from other sectors (especially in relation to carbon dioxide), and thus the mitigation pathways also need to be different. The overall sustainability challenges related to food production, and possible solutions, have to be considered while developing CNRFS (UNFCCC, 2018; FAO, 2019). Climate change mitigation within the agricultural sector can be addressed effectively if local and national level actions are accompanied by international and regional level cooperation and sharing of knowledge. In this context, international collaborative research and the related co-learning have to be supported at all levels. This book focuses on the unique opportunity to cooperate and bring together experts from Asia, Australia, Europe, and Africa to share their experiences with CNRFS that can potentially contribute to GHG emission reduction and carbon sequestration (including carbon storage in agricultural soils) and, at the same time, foster climate resilience and contribute to overall sustainability. Regarding the implications for farmer income and livelihoods, food sovereignty cannot be ignored in this context, since success depends on the farm-level implementation of CNRFS-proposed innovations and the benefits gained.

The agriculture sector as a potential solution

Major changes in the agriculture sector would be required if the impact of climate change is to be limited to 1.5°C. Overcoming the challenges to reduce GHGs in the agriculture sector will require technological, investment, and policy solutions. These solutions will need to be targeted at the global as well as specific scales, at developed and developing countries, and at large- and small-scale agriculture, and should be environmental friendly, socially acceptable, and economically feasible.

Technological solutions

It is technically feasible for agriculture to become close to carbon neutral, relying on supply-side mitigation measures alone, although this depends on optimistic assumptions about the potential of soil carbon sequestration (SCS). Based on full deployment of available emission reductions, coupled with carbon sequestration opportunities, the global technical mitigation potential of the agricultural sector in 2030 is estimated to be 5,500 to 6,000 $MtCO_2eq\ yr^1$ (Smith, 2016).

Global versus local scales

Targets are usually set at global or national levels, putting together the required emission reductions for various sectors. It has been suggested that the first step in reducing emissions from agriculture is to produce food as efficiently as possible, i.e. *changing how we farm* (McKinsey & Company, 2020). The report found that a set of proven GHG-efficient farming technologies and practices (which are already being deployed) could achieve about 20% of the sector's required emission reduction by 2050. The report found that top 15 mitigation measures could deliver about 85% of the total 4.6 GtCO$_2$e mitigation potential from GHG-efficient farming by 2050, compared with business-as-usual emissions. These measures included reducing GHG emissions in farm machinery (zero-emissions on-farm machinery), livestock sector (breeding for reduced GHG emissions, improving animal health, optimizing livestock feed, and increasing production efficiency), crop management (direct seeding in rice), water management (improve paddy rice water management), fertilizer management (improve nitrogen fertilization strategies in rice, nitrification inhibitors, expand slow release fertilizers), and soil management (minimum tillage and/or no-tillage practices). The various chapters in this book focus on some of these selected GHG-reducing measures in the farming systems.

According to McKinsey & Company (2020), there *is also a need to change our dietary patterns, i.e. what we eat, and reduce food wastage.* Overall, achieving emissions-reduction goals is not possible without billions of people pro-actively changing their diets by eating less beef and lamb meat. There is room for improvement here, as the current average global consumption of ruminant animal protein (mainly beef and lamb) is three times the recommended level, and almost one-third of all food produced in the world is wasted. Reducing food waste and shifting to plant-based protein diets will be beneficial to both the environment and human health (Tziva et al., 2020).

Developed versus developing countries

In general, developing countries are the largest and fastest growing source of GHG emissions. Between 1990 and 2014, they were largely responsible for the 15% increase in global non-CO$_2$ emissions from agriculture (Blandford and Hassapoyannes, 2018), whereas OECD countries (the Organisation for Economic Co-operation and Development, OECD) as a whole experienced a slight reduction in non-CO$_2$ emissions over the same period. Improvements in production efficiency have contributed to this reduction by lowering the emission intensity of agricultural output (MacLeod et al., 2015). However, the rate of decline in intensity appears to be slowing (OECD, 2021). Reduced deforestation rates and increased afforestation in several regions of the world have resulted in net CO$_2$ emissions from forestry and other land uses, falling in both developed and developing countries (Blandford and Hassapoyannes, 2018; IPCC, 2019; Smith et al., 2016).

Reducing GHG emissions must also be considered in relation to food security issues, particularly in the developing world. A useful entry point for mitigation would be a focus on reducing emissions intensity by addressing productivity improvements and yield gaps (IPCC, 2019). This approach can simultaneously meet food security, rural development, and climate change mitigation goals (Vermeulen et al., 2012; Wollenberg, 2017). The extent to which reducing emissions intensity will result in reductions of absolute emissions depends on changes in total production (Leahy, 2020). Absolute emissions will generally rise if increased productivity is used to generate more food to meet nutritional or economic goals, and reducing emissions intensity will only reduce absolute emissions if total production does not increase at a faster rate (Gerber et al., 2013). Recent studies suggest that agricultural land could expand further in the absence of changes to technological progress, dietary patterns, food distribution, and markets (Roos et al., 2017).

Some technological solutions will require proactive strategies for implementation in the developing world. For example, disruptive change from novel mitigation technologies, such as methane vaccines and methane and nitrous oxide inhibitors, could produce significant global benefits, but special effort needs to be given to adoption pathways in developing countries. Development of methane inhibitors is largely being carried out in developed countries, with the most advanced product (3-NOP) so far showing high efficacy (>30% reduction) in feedlot systems (Hristov et al., 2015), with limited suitability for grazing-based systems (Reisinger et al., 2018).

Approximately one-third of all food produced is never consumed. Food loss takes place early in the supply chain during production, transportation, and storage, particularly in developing countries where losses are driven by lack of access to technology and cold-storage infrastructure. In developed countries, food waste more likely occurs at the retail and consumption stages and is prevalent in higher-income regions (Nicastro and Carillo, 2021). To meet a 1.5°C target, food loss and waste would need to fall from about 33% in recent years to under 30% by 2030, and 20% by 2050. This would result in reducing overall emissions from food waste by about 40% globally. Large-scale awareness programmes, educating youth and fixing accountability, is the way to go forward to reduce food wastage.

Large scale versus small scale

New farm practices and technologies need to be relevant to small-scale farms around the world, since most farms in developing countries are small. In fact, farms of two hectares or smaller produce 30–34% of the global food supply and account for about 75% of farms (Lowder et al., 2016). The pace of change in agriculture is slow, partly due to this fragmentation, particularly when it comes to adoption of new technologies (Gandhi et al., 2016).

Investment solutions

The private sector is driving policy around reducing GHG emissions in the agricultural sector, with governments struggling to keep up to some extent. Governments are potentially being outflanked by targets and requirements set by large international food companies. The on-farm supply chain emissions of such companies can often account for significant proportions of their total GHG footprint (Leahy et al., 2020). For example, 57% of Danone's "scope 3" GHG emissions are related to the purchase of agricultural products such as milk (Danone, 2017). Many international food and beverage companies (e.g. Danone, Mars Inc., Nestle, Tesco, Coca-Cola Co., Kellogg, PepsiCo., Unilever PLC) are driving climate goals (Leahy et al., 2020). These companies are setting ambitious emissions targets and increasingly mandating that their suppliers also provide a product that meets the company's stated climate agenda.

Considering the extent of emissions from livestock industries, it is critical that industries actively drive the adoption of climate-smart policies and invest significantly. Total GHG emissions from livestock supply chains alone are estimated to be around 7.1 Gt CO_2eq/yr (Gerber et al., 2013). The GHG emissions of 35 of the world's largest meat and dairy companies are reported to account for up to 1 Gt CO_2eq/yr (14%). The on-farm supply chains from these companies are a major source of emissions (GRAIN and the Institute for Agriculture and Trade Policy, 2018). It is likely that these company goals, coupled with global market dynamics, will increasingly shape production systems of the future. While this approach may influence internationally traded products, it may have limited impact on subsistence and smallholder farmers that provide more than half of total food production in many developing countries (Rapsomanikis, 2015). Investment, development, commercialization, and scaling of next-horizon technologies should greatly accelerate efforts to reduce GHG emissions in the agriculture sector.

There are a range of promising technologies at various stages of development that could have significant GHG abatement potential in the crop and livestock sectors. These include gene editing for disease resistance or for enhanced carbon sequestration, plant and soil microbiome technology, aerobic rice, direct methane capture from beef and dairy cattle, perennial row crops, inhibition of enteric fermentation through vaccines, and novel feed additives (Al-Azzawi, 2021, IGI, 2021; ITIF, 2020; McKinsey & Company, 2020).

Policy solutions

To achieve any given level of mitigation in GHG emissions at minimum economic cost, two requirements are necessary for policy measures (OECD, 2021). The first requirement is the use of market-based policy instruments that achieve a common price for GHG emissions (such as an emissions tax or emissions trading scheme). The second is that coverage of the market-based policy includes the largest possible share of global emissions from all regions and sectors. These two policy requirements should ensure that the lowest cost mitigation measures are adopted, given the large heterogeneity in marginal abatement costs among agents, sectors, and regions.

So far, no single country has set a mandatory carbon price for agricultural emissions and current evidence suggests considerable reluctance to applying other climate policies with comparable stringency to agriculture (Leahy et al., 2020). A recent review on agricultural GHG mitigation pathways stated that a more realistic view is needed if we are to avoid modelled emission scenarios providing an overly optimistic picture of mitigation potentials from the agricultural sector (Leahy et al., 2020). While there are entry points for mitigation of agricultural GHGs outside government price policies, many questions remain unanswered around their efficacy and scalability, requiring a concerted effort to bridge the gap from modelled emissions to realistic policy pathways.

Integrated policy interventions that span supply and demand approaches will be required to achieve agricultural mitigation pathways that are aligned with the 1.5°C pathway (IPCC, 2019). Agricultural trade is subject to a wide range of constraints and distortionary subsidies that reflect powerful special interests. Furthermore, developing countries' desire for food self-sufficiency and protection from food price spikes must be considered. It is worth noting that some of these spikes have been linked to increased biofuel demand driven by climate policies in the energy sector of developed countries (Anderson, 2016). Hence, international coordination is fundamental to addressing concerns about competitiveness, ensuring environmentally effective outcomes, and avoiding negative consequences at the trans-national scale (Blandford and Hassapoyannes, 2018).

Current evidence suggests reluctance to apply rigorous climate policies to agriculture, even in developed countries. In theory, substantial reductions in agricultural emissions could be attained through a number of mechanisms, including the widespread introduction of price-based policies or other measures with an implicit price (Leahy et al., 2020). However, there appears to be little current interest in such strategies. For example, New Zealand is the only country actively considering a compulsory price on agricultural emissions, although more than 100 countries have included agriculture mitigation in their NDCs (Richards, 2019).

Global versus local scales

The global desire to reduce GHG emissions in agriculture is currently weak and needs to be strengthened or the lack of progress will stifle efforts to meet the goals of the Paris Agreement to limit global warming to 1.5°C or well below 2°C (OECD, 2021). Only 38% of agriculture emissions are covered by nationally determined commitments under the Paris Agreement (Hönle et al., 2019). This shows that governments are not prioritizing actions to cut emissions and combat climate change. However, policy solutions do exist, despite the reality of barriers to policy implementation. Solutions include choosing policy options that can navigate trade-offs in economic impacts between different interest groups. Other options could address the practical challenges and transaction costs related to measuring, reporting, and verifying the extent of reductions in GHG emissions.

To achieve a 1.5°C pathway, 6–8 Gt of carbon dioxide sequestration is required. If all of this was to be delivered through forestry, it would require reforesting 50–60% of the total area that has been deforested over the past 150 years.

However, there are other options for enhancing soil carbon through regenerative agricultural practices such as low- and no-till agriculture, green manuring, composting, cover crops or crop rotations, and legumes sown in pastures (Raphaela, 2016).

Developed versus developing countries

It is estimated that smallholder farming contributes about 1.7 Gt CO_2eq/yr emissions (Vermeulen and Wollenberg, 2017). Given the concerns about food security, self-sufficiency, food sovereignty, and rural poverty, most developing countries will find price-based policies or stringent regulatory policies targeting agricultural GHG emissions even harder to implement than developed countries (IPCC, 2019). Therefore, implementation of policies to reduce GHG emissions from the agriculture sector is highly challenging for both developed and developing countries, although there are different constraints, depending on the context.

Modelling studies support this view (Hasegawa et al., 2018), concluding that climate change mitigation actions could have potentially adverse side effects on food security for some populations. More nuanced implementation and targeted support mechanisms for vulnerable groups could overcome many of the negative consequences arising from blunt price-based policies (Fujimori et al., 2018; Loboguerrero et al., 2019). However, the question remains as to whether finance, governance, and institutional capacity exist, together with the political will, to deliver policy arrangements of such complexity at the necessary scales required (Grewer et al., 2018).

Climate action and sustainable development goals

The sustainable development goal (SDG) 13 is about climate action and it is one of the United Nations 17 SDGs (United Nations, 2020). The SDGs provide economic and political legitimacy to introduce sustainable initiatives. Addressing climate action must be seen as an opportunity by countries and not merely an obligation. At the same time, climate action will address other SDGs and, in the process, support global consensus and fulfil national commitments to sustainably manage the Earth's resources. Countries rebuilding economies after COVID-19 should seize this occasion to include green solutions or green growth strategies that can also address climate change. If carefully planned and implemented, actions could simultaneously address SDG 1 (No poverty), SDG 2 (Zero Hunger), and SDG 13. A good example could be to introduce direct-seeded rice (DSR) to reduce methane emissions in areas where transplanted paddy rice is a dominant system, without compromising on productivity. In dry DSR systems with a different water regime, and less anaerobic conditions, methane emissions could be substantially reduced compared to transplanted paddy rice, due to lower CH_4 production and release (Li et al., 2019). Thus, smallholder rice farmers practising DSR will be able to realize good yields (SDG 2) and, at the same time, contribute to methane reductions from rice fields (SDG 13). A sustainable food system incorporating improvements at all levels of the value chain from production to consumption, in other words a green transition from farm to fork, is the way forward. Implementation of the commitments made in the Paris Agreement to combat the climate crisis is essential for the achievement of the SDGs.

Innovative and systemic solutions: climate-neutral and resilient farming systems

Given the huge challenges facing farming systems in general (*soil and water pollution and degradation, biodiversity decline, emerging pests and diseases, GHG emissions, etc.*) and conventional systems such as transplanted paddy rice (*low water and nutrient use efficiency, high GHG emissions*), as well as other unforeseen shocks and risks, there is an increasing need for systematically identifying, demonstrating, implementing, and assessing/evaluating innovative solutions that are sustainable and reduce GHGs.

Need for assessing current systems and developing appropriate CNRFS

A transition to sustainable CNRFS can be achieved, first, by obtaining a better understanding of current systems, including their strengths and weaknesses at field/landscape scale, their dynamics, interdependencies, local knowledge, drivers and barriers, as well as existing resources and institutions. Adopting a systems approach is the way to better understand and synthesize the complexity of existing farming systems in particular regions in a stepwise manner by: (i) analysing the problems, root causes, opportunities to improvements; (ii) examining technological, social/policy, and economic factors (e.g. market constraints and consumer's preferences to assess the opportunities and the willingness to pay for the products); and (iii) developing pathways to transform the existing value chains towards more economically, socially, and environmentally sustainable ones.

At the same time, attention must be paid to environmental conditions including functional agri-biodiversity and provision of ecosystem services, damage due to pests, diseases and resilience against weather extremes. In order to ensure farmer adoption of CNRFS, it will be necessary to consider the farm household resource-use patterns for on-farm and off-farm activities, the cooperation along agro-food value chains, the embeddedness of farming systems and practices in agro-ecological settings, economic opportunities (and limitations), institutional context, and cultural values. The related data should ensure that interdependencies, dynamics, and farmer or local knowledge, as well as existing resources and institutions, are duly considered while designing CNRFS.

The participatory diagnosis and co-analysis of given farming systems should include quantitative and qualitative methodologies and tools. The analysis must make use of and combine concepts and approaches from the natural sciences and social sciences, including quantitative meta-analyses of available data. Good baseline data and indicators derived from the analysis would be necessary for measuring and monitoring progress observed with the introduction of CNRFS.

Co-designing and promoting CNRFS

After taking inputs from the initial assessment of current systems, the next step should be co-designing combinations of the most effective solutions that are suitable for particular agro-ecological settings that can lead to improvements (Dainese et al., 2019). An important advancement would involve combining nature-based and technology-based solutions supported by the right institutional and policy-based approaches. Such a combination, where simple and easy to adopt local- and nature-based solutions in many settings could, in fact, help in carbon assimilation, as well as reducing GHGs (EEA, 2021). Emerging evidence indicates that managing whole landscapes has tremendous potential for translational changes in sustainability and CNRFS (Martin et al., 2019). Landscape factors that promote the richness and abundance of functional groups, such as natural enemies and pollinators, enhance the provision of multiple ecosystem services and contribute to crop yield quantity and quality (Martin et al., 2019, Dainese et al., 2019). First, one must prioritize to upscale proven local solutions. Second, the aim should be to combine upscaling of local solutions with management of landscapes. These include (i) the restoration of green infrastructure, i.e. perennial near-natural or semi-natural habitats that act as sources for beneficial organisms such as crop pollinators and arthropod and vertebrate predators of pest insects (Beddington et al., 2002), and (ii) the management of crop diversity. Both factors have been shown to improve natural pest control (Martin et al., 2019).

The CNRFS chosen should be sustainable in *environmental* terms (in particular contributing to CNRFS), in *economic* terms (e.g. relatively low cost and cost-efficient, contribute to farmer income), and in *social* terms (e.g. fostering diverse diets, minimizing health risks, appropriate to the particular socio-cultural context).

Only if farmers and agri-food chain actors and stakeholders can gain from the new CNRFS will they be motivated to actually adopt and use the solutions? The sustainable solutions need to address the needs in the *pre-production phase* (**inputs**: seeds, nutrients, water, crop and livestock management), as well as in the *post-production phase* (**outputs**: *grains, bio-residues, storage*), and market preferences (e.g. consumer choices, increasing demand for healthy diets). However, for successful adoption and upscaling, the innovative systemic solutions require enabling institutions and policy support (Nagothu, 2015). The systemic context shaping farmer's/stakeholder's ability to implement these solutions is depicted in Figure 1.1, where a transition from the current system towards CNRFS involves interventions at different levels.

In order to design promising CNRFS pathways that are appropriate to the different agro-ecological zones, particular attention should be paid to replacing external farm inputs (i.e. fertilizers, pesticides) with internal farm inputs (i.e.

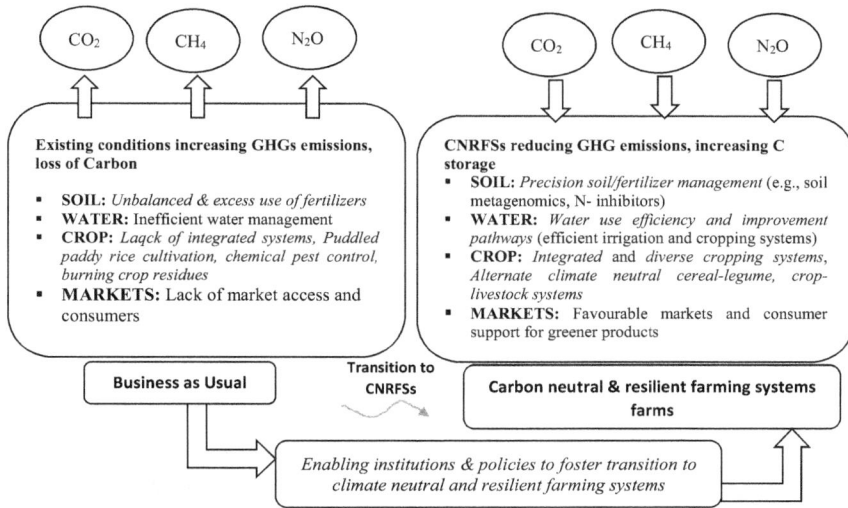

Figure 1.1 Conceptual diagram showing pathways leading towards CNRFS.
Source: Authors' own design.

nature-based solution such as biopesticides/biocontrol, biological solutions, improved soil services, conservation agriculture, organic farming). The key would be to increase both sustainability and resilience against future climate stresses and extremes. Examples of improved management practices that will be further discussed in subsequent chapters include the following:

- *Soil/nutrient management using soil metagenomics*: An important area, which has the potential for GHG reduction if precise management is applied. This can be enhanced by innovative information and communication tools/sensors/mobile apps for upscaling and improving precision use of farm inputs (Chapter 2).
- *Water management*: Improving water use efficiency/water productivity (e.g. alternate wetting and drying/drip irrigation, conservation agriculture) that can significantly reduce GHGs (Chapter 3).
- *Integrated pest, weed, disease management*: Using nature-based solutions, e.g. through enhanced farm and landscape functional diversity, integrating physical and other non-chemical measures to manage pests and diseases (Chapter 4).
- *Crop management*: Improving efficiency through systems such as climate resilient rice systems/agro-ecological farming practices, short duration flood and drought tolerant crop varieties and cover/catch crops. Here it is also important to consider specific genetic solutions for adaptation to a particular environment or climate while developing the new crop management practices (Chapters 5–8).

- *Post-harvest, processing, marketing/distribution to consumption*: Key questions relate to consumer preferences, the sharing of higher production costs along the value chain (including the shares that consumers are willing to cover), addressing fluctuations in market prices and feedback mechanisms on products and production systems (Chapter 9).

Irrespective of farm size, ecologically sound practices (e.g. minimum tillage, legume N-fixing crops, stubble retention) that can reduce environmental impacts, thereby enhancing climate neutrality and keeping food production systems in safe spaces, should be introduced (Bommarco et al., 2013, Dainese et al., 2019). Further, combining local measures with landscape management concepts (through permanent green infrastructure, enhanced crop diversity, and coordinated placement of agri-environment schemes) provides a novel pathway to more sustainable agriculture (Martin et al., 2019).

Digital and space-based technologies represent another line of promising and emerging solutions to counter environmental costs of crop production systems, improve efficiency, and enhance climate resilience (King, 2017). Such solutions include precision farming with threshold-based and spatially targeted application of pesticides and fertilizers, more efficient irrigation systems, and sensor- and remote sensing-based monitoring of crop growth and potential risks (King, 2017). A key factor to bridge gaps between scientific theoretical knowledge and practical implementation is the continuous involvement, training, and co-design of solutions with farmer's communities and other stakeholders (Kleijn et al., 2019). Combining the nature- and technology-based components to pilot and upscale systematically designed innovative solutions will be the way forward in the future. Thus, it should be possible to overcome the limitations and risks of current conventional farming systems, including stagnation of yields, increasing yield losses due to pests and extreme weather events, degradation of soils, and emission of GHGs.

Institutional changes involve a whole range of factors such as implementation of conducive policies, enabling environments and effective value chains (Glover et al., 2019). Also the influence of risk and uncertainty in relation to socio-economic and marketing constraints will be important to include while implementing new systems (Reardon et al., 2019). The diffusion of CNRFS will depend on the role of service providers (e.g. business, advisory, information, extension) to a large extent. The transformation process should adopt a wider stakeholder perspective in order to come up with strategies for fostering collaboration among actors and enhancing uptake of innovations (FAO, 2014). The overall aim is to reach net-zero emissions, together with reduced external inputs and more stable yields that can contribute to sustainability along agri-food chains. An important step in the process is to address the economic, social, and environmental sustainability challenges related to current farming systems and develop measurable baseline indicators, some of which have already been addressed in the previous stages. Attention is needed to reduce GHG emission in the process.

Multi-actor partnerships to enhance CNRFS

An inclusive multi-actor approach aims at a more demand-driven innovation process through the genuine involvement of diverse actors all along the project and different segments of the agri-food chains, from farm to fork. Multi-actor platforms (MAPs) have the potential to bring diverse actors together in a structured process of interactive learning, sharing, empowerment, and collaborative governance (Brouwer et al., 2015). Together, the actors can discuss opportunities and the ways to achieve a desired set of goals. The MAPs can promote innovation in the face of complexity, uncertainty and risk, and strengthen science-policy linkage. This is achieved by building trust and continuous dialogue among the actors with interconnected but potentially divergent interests or viewpoints (Brouwer et al., 2015). A multi-actor approach involves working directly with farmers, managers, civil society groups (e.g. youth, indigenous groups, and women); non-governmental organizations (NGOs); and scientific, policy, and business communities (EIP-AGRI, 2020). Their involvement in analysis, co-design, piloting, and upscaling of promising CNRFS will be highly relevant for promoting new CNRFS. The knowledge and experience of key agri-food actors combined with scientific knowledge will help to develop applicable best practices. Setting up sustainable MAPs, however, is a significant challenge (Reid et al., 2014), requiring a whole range of skills, support, structure, and process.

The regional and sectoral needs and contexts (environmental, socio-economic, geographical, cultural) must be considered from an early stage in the process of transition to CNRFS. The process should ensure that all relevant food systems stakeholders are actively engaged so that:

- potential improvements to existing food, farming, and cropping systems can be jointly identified together with all stakeholders engaged in the MAPs. The cross-sectoral representation in MAPs and use of systemic solutions will ensure that co-created solutions will go far beyond mere technological innovations and that a cross-sectoral agri-food chain (markets, consumer, demand) perspective is applied (European Commission, 2017).
- it becomes easier to build on the existing MAPs (including youth and women organizations, government agencies, small medium enterprises (SMEs), producers, processors, retailers, food service providers, consumers), rather than creating new ones. Comparable structures comprised of representatives from producer associations, farmer innovation/learning circles, and supply chain initiatives, already identified, should be engaged (Nagothu et al., 2018).

In line with the multi-actor approach, defining and prioritizing the research needs together with the MAPs should be done simultaneously, including piloting the most appropriate combinations of solutions to accelerate the transition to CNRFS (Nagothu et al., 2018). Specific care should be taken to engage young professionals (e.g. young farmers, young fishers, young researchers, young entrepreneurs), SMEs, consumers, and citizens.

Stakeholder perceptions about climate change

While drafting this book chapter, farmers, scientists, and government agencies in some of the rice growing regions in Piedmont, northern Italy, and the Odisha and Assam provinces in India, were contacted during September-October 2021. The purpose was to seek their opinion and perceptions on the current climate crisis and associated vulnerability and the implications of the Sixth IPCC report. In general, stakeholders in both India and Italy perceived the climate crisis as an immediate threat – exhibited by temperature changes, i.e. long dry periods alternating with unexpected intensive rainfall, warmer in the winter months and during early spring. Farmers, whether in Italy or in India, viewed that changes to climate will have a serious and direct influence on food production due to extreme climate variability. According to the respondents, farmers will be one of the communities most affected by the climate crisis because it is difficult to plan and grow crops in highly variable environments.

During the discussions, farmers and government agencies suggested combating climate crisis with climate-neutral solutions in agriculture that can create maximum impact – and able to address both adaptation and mitigation simultaneously (to reduce GHGs and store carbon in the soils). Some of the measures suggested were (i) minimum tillage of the soil, (ii) growing climate resilient crop varieties with short duration (to reduce GHGs and fix soil carbon), (iii) diverse crops, cereal-legume rotations, and (iv) mulching with biomass to increase organic matter in the soils and enhance soil health. Wherever possible, farmers were of the opinion that these practices should be combined with agroforestry to increase agri-biodiversity. One farmer practising organic rice farming in Piedmont suggested that incentives should be given for regenerative organic agriculture, training agricultural technicians who could, in turn, assist farmers in the green transition movement. The farmers also expressed that exploitation of alternative energies such as solar and wind in the agriculture sector should be explored as they contribute to reduction of GHG emissions.

Although there are incentives or subsidies, they are not directly given to farmers or to support actions that can help in combating the climate crisis. Overall, the farmer and other stakeholders' perception was that investments for scaling up climate-neutral agricultural technologies that can reduce GHG emissions are going to be a challenge. Farmers in the two countries also expressed concerns that current agricultural insurance programmes do not cover crop damages and losses due to climate extremes.

During the interviews, stakeholders expressed that in Europe it may be possible to tap the European funds for the Regional Rural Development Plan for supporting farmers to reduce GHG emissions (European Commission, 2017). The current Horizon Europe programme provides an opportunity for scientists and stakeholders across the EU to cooperate on research and development to develop carbon-neutral and green technologies (European Commission, 2021). Whereas in countries such as India, although subsidies exist in the farming sector, it will be challenging to access funding for reducing emissions from agriculture, as the country priorities are different compared to the EU region. It can be worse in African regions, where funds are even more limited.

The EU is in the forefront when it comes to policy support (EU level and national) to address the climate crisis. In irrigated regions, it will be important to make regulations for water use and irrigation and improve efficiency. Farmers during the interviews expressed that access to climate-neutral technologies, bio-based solutions, and inputs are important to reduce GHG emissions from the agriculture sector.

Outline of the book

This book is divided into different thematic chapters with a common objective and at the same time cover cross-cutting issues related to technology, nature-based solutions, socio-economic and policy perspectives relevant for upscaling the CNRFS.

The first chapter provides a comprehensive review of the concept of climate-neutral farming systems, main sources of GHG emissions, risks, and transition pathways to develop and promote CNRFS that can contribute to relevant SDGs (especially SDGs 2, 13, and 15) and policy implications. This is followed by several chapters demonstrating promising CNRFS in rice and other food crop systems, based on experiences from different regions/countries including South and SE Asia, Europe, and Africa.

A separate chapter discusses the relevance of value chain analysis and integrating cross-cutting issues, including stakeholder engagement and multi-actor partnerships, and market and policy perspectives relevant for upscaling the CNRFS. Towards the end, the last chapter summarizes the key messages and lessons learnt from each chapter with specific policy recommendations and framework conditions necessary to be put in place to upscale the innovative CNRFS at a systemic level.

Conclusions

Achieving the major changes required to sufficiently reduce GHG emissions to meet the necessary targets may be more challenging for agriculture than for other sectors. In addition, the agriculture sector has a number of other complex objectives to consider alongside climate goals, including food and nutritional security, biodiversity, and the livelihood of farmers and farming communities.

Rather than just dwelling on how we produce our food, we must change the way and what we eat, how we reduce food wastage, how we manage our forests and carbon sinks, and how we apply next-horizon technologies. But we need to act swiftly; otherwise, emissions in agriculture will continue to grow and contribute to heating the planet to dangerous levels.

However, we must not lose hope. Agriculture has responded to humanity's greatest challenges throughout the course of human history, and there is no reason why the current challenges cannot also be addressed. As evidence of this, in the past 50 years the agriculture sector has increased food production to a level that many believed impossible. During this crucial window for global action on climate change, the sector now has another opportunity to make a major contribution to humanity's success.

References

Al-Azzawi, M., Bowtell, L., Hancock, K. and Preston, S. (2021) 'Addition of Activated Carbon into a Cattle Diet to Mitigate GHG Emissions and Improve Production', *Sustainability*, vol. 13(15), p. 8254.

Anderson, K. (2016) 'International Food Price Spikes and Temporary Trade Policy Responses', Palgrave Studies in Agricultural Economics and Food Policy. In: *Agricultural Trade, Policy Reforms, and Global Food Security*, pp. 177–206. Palgrave Macmillan, New York.

Beddington, J., Asaduzzaman, M., Clark, M., Fernandez, A., Guillou, M., Jahn, M., Erda, L., Benedict, M. A. and McMahon, E. T. (2002) 'Green Infrastructure: Smart Conservation for 21st Century', *Renewable Resources Journal*, vol. 20, pp. 12–17.

Blandford, D. and Hassapoyannes, K. (2018) 'The Role of Agriculture in Global GHG Mitigation', Available at: https://dx.doi.org/10.1787/da017ae2-en

Boetzl, F. A., Schuele, M., Krauss, J. and Steffan-Dewenter, I. (2020) 'Pest Control Potential of Adjacent Agri-Environment Schemes', *Journal of Applied Ecology*, vol. 57, pp. 1482–1493.

Brouwer, H., Jim Woodhill, J., Hemmati, M., Verhoosel, K. and van Vugt, S. (2015) 'The MSP Guide, How to Design and Facilitate Multi-stakeholder Partnerships', The Centre for Development of Wageningen UR, Netherlands, ISBN 978-94-6257-542-4.

Bommarco, R., Kleijn, D. and Potts, S. G. (2013) 'Ecological Intensification: Harnessing Ecosystem Services for Food Security', *Trends in Ecological Evolution*, vol. 28, pp. 230–238.

C2ES (2021) 'Global Emissions'. Available at: https://www.c2es.org/content/international-emissions/

Dainese, M., Martin, E. A., Aizen, M. A., Albrecht, M., Bartomeus, I., Bommarco, R. and Steffan-Dewenter, I. (2019) 'A Global Synthesis Reveals Biodiversity-mediated Benefits for Crop Production', *Science Advances*, vol. 5(10), pp. 121.

Danone (2017) 'Environmental Performance'. Available at: http://iar2017.danone.com/performance-in-2017/key-performance-indicators/environmental-performance/

Deutsch, C. A, Tewksbury, J. J., Tigchelaar, M., Battisti, D. S, Merrill, S. C., Huey, R. B. and Naylor, R. L. (2018) 'Increase in Crop Losses to Insect Pests in a Warming Climate', *Science*, vol. 361, pp. 916–919.

European Commission (2017) 'Rural Development Programs by Country'. Available at: https://ec.europa.eu/info/food-farming-fisheries/key-policies/common-agricultural-policy/rural-development/country_en

European Commission (2021) 'Horizon Europe'. Available at: https://ec.europa.eu/info/research-and-innovation/funding/funding-opportunities/funding-programmes-and-open-calls/horizon-europe_en

EEA (European Environment Agency) (2021) 'Nature-based Solutions in Europe: Policy, Knowledge and Practice for Climate Change Adaptation and Disaster Risk Reduction'. Available at: https://publications.europa.eu/en/publications

EIP-AGRI (2020) 'EIP-AGRI Brochure Horizon 2020 Multi-actor Projects'. Available at: https://ec.europa.eu/eip/agriculture/en/publications/eip-agri-brochure-horizon-2020-multi-actor

FAO (Food and Agriculture Organization) (2013) 'FAOSTAT Emissions Database'. Available at: http://faostat.fao.org/

FAO (Food and Agriculture Organization) (2014) *Developing Sustainable Food Value Chain: Guiding Principles*. Rome: FAO.

FAO (Food and Agriculture Organization) (2015) 'Climate Change and Food Security: Risks and Responses'. Available at: https://www.fao.org/3/i5188e/I5188E.pdf

FAO (Food and Agriculture Organization) (2018) *The State of Food Security and Nutrition in the World*. Rome: FAO. Available at: http://www.fao.org/3/I9553EN/i9553en.pdf

FAO (Food and Agriculture Organization) (2019) *FAOSTAT*. Rome: FAO. Available at: http://www.fao.org/faostat/en/#home

FAOSTAT (2020) *World Food and Agriculture: Statistical Year Book 2020*. Rome: FAO. Available at: https://www.fao.org/documents/card/en/c/cb1329en/

Foley, J. A., Ramankutty, N., Brauman, K. A., Cassidy, E. S., Gerber, J. S., Johnston, M. and Zaks, D. P. (2011) 'Solutions for a Cultivated Planet', *Nature*, vol. 478(7369), pp. 337–342.

Fujimori, S., Hasegawa, T., Rogelj, J., Su, X., Havlik, P. and Krey, V. (2018) 'Inclusive Climate Change Mitigation and Food Security Policy Under 1.5°C Climate Goal', *Environmental Research Letters*, vol. 13, p.074033.

Gandhi, P., Khanna, S. and Ramaswamy, S. (2016) 'Which Industries Are the Most Digital (and Why)?'. Available at: https://hbr.org/2016/04/a-chart-that-shows-which-industries-are-the-most-digital-and-why

GDACS (2021) 'Overall Green Alert Drought for Southern Canada and Northern USA-2021'. Available at: https://www.gdacs.org/media.aspx?eventtype=DR&eventid=1014320

Gerber, P. J., Steinfeld, H., Henderson, B., Mottet, A., Opio, C., Dijkman, J., Falcucci, A. and Tempio, G. (2013) *Tackling Climate Change through Livestock: A Global Assessment of Emissions and Mitigation Opportunities*. Rome: Food and Agriculture Organization of the United Nations (FAO).

Glover, D., Sumberg, J., Ton, G., Andersson, J. and Badstue, L. (2019) 'Rethinking Technological Change in Small and/or Large Scale Farmers', *Agriculture Outlook*, vol. 48(3), pp. 169–180.

GRAIN and the Institute for Agriculture and Trade Policy (2018) 'Emissions Impossible: How BigMeat and Dairy Are Heating Up the Planet'. Available at: https://www.grain.org/article/entries/5976-emissions-impossible-how-bigmeat-and-dairy-are-heating-up-the-planet

Grewer, U., Nash, J., Gurwick, N., Bockel, L., Galford, G., Richards, M., Junior, CC., White, J., Pirolli, G. and Wollenberg, E. (2018) 'Analyzing the Greenhouse Gas Impact Potential of Smallholder Development Actions Across a Global Food Security Program', *Environment Research. Letters*, vol. 13, p. 044003.

Myhre, G., Shindell, D., Bréon, F.-M., Collins, W., Fuglestvedt, J., Huang, J., Koch, D., Lamarque, J.-F., Lee, D., Mendoza, B., Nakajima, T., Robock, A., Stephens, G., Takemura, T. and Zhang, H. (2013) 'Anthropogenic and Natural Radiative Forcing'.

In: *Climate Change 2013': The Physical Science Basis. Contribution of Working Group I to the Fifth Assessment Report of the Intergovernmental Panel on Climate Change* [Stocker, T.F., D. Qin, G.-K. Plattner, M. Tignor, S.K. Allen, J. Boschung, A. Nauels, Y. Xia, V. Bex and P.M. Midgley (eds.)]. Cambridge, UK, and New York, NY: Cambridge University Press.

Hasegawa, T., Fujimori, S., Havlík, P., Valin, H., Bodirsky, B. L., Doelman, J. C., Fellmann, T., Kyle, P., Koopman, J. F. L., Lotze-Campen, H., Mason-D'Croz, D., Ochi, Y., Domínguez, I. P., Stehfest, E., Sulser, T. B., Tabeau, A., Takahashi, K., Takakura, J., van Meijl, H., van Zeist, W.-J., Wiebe, K., and Witzke, P. (2018) 'Risk of Increased Food Insecurity Under Stringent Global Climate Change Mitigation Policy', *Nature Climate Change*, vol. 8, pp. 699–703.

Hönle, S. E., Heidecke, C. and Osterburg, B. (2019) 'Climate Change Mitigation Strategies for Agriculture: An Analysis of Nationally Determined Contributions, Biennial Reports and Biennial Update Reports', *Climate Policy*, vol. 19(6), pp. 688–702.

Hristov, A. N., Oh, J., Giallongo, F., Frederick, T. Q., Harper, M. T., Weeks, H. L., Branco, A. F., Moate, P. J., Deighton, M. H., Williams, S. R. O., Kindermann, M. and Duval, S. (2015) 'An Inhibitor Persistently Decreased Enteric Methane Emission From Dairy Cows With No Negative Effect on Milk Production', *Proceedings of National Academy of Science*, vol. 112, pp. 10663–10668.

IGI (2021) *IGI Launches New Research in Net-Zero Farming and Carbon Capture.* Available at: https://innovativegenomics.org/news/net-zero-farming-carbon-capture/

IPCC (Intergovernmental Panel on Climate Change) (2018) 'Global Warming of 1.5°C, å'. Available at: https://www.ipcc.ch/sr15/

IPCC (Intergovernmental Panel on Climate Change) (2019) 'IPCC Special Report on Climate Change, Desertification, Land Degradation, Sustainable Land Management, Food Security, and Greenhouse Gas Fluxes in Terrestrial Ecosystems: Summary for Policymakers'. Available at: https://www.ipcc.ch/srccl/

IPCC (Intergovernmental Panel on Climate Change) (2021a) 'Climate Change, Widespread, Rapid and Intensifying'. Available at: https://www.ipcc.ch/2021/08/09/ar6-wg1-20210809-pr/

IPCC (Intergovernmental Panel on Climate Change) (2021b) 'Intergovernmental Panel on Climate Change Sixth Assessment Report'. Available at: https://www.ipcc.ch/report/ar6/wg1/

ITIF (2020) 'Gene Editing for the Climate: Biological Solutions for Curbing Greenhouse Emissions'. Available at: https://www2.itif.org/2020-gene-edited-climate-solutions.pdf

King, A. (2017) 'The Future of Agriculture: A Technological Revolution in Farming Led by Advances in Robotics and Sensing Technologies Looks Set to Disrupt Modern Practice', *Nature*, vol. 544, p. s23.

Kleijn, D., Bommarco, R., Fijen, T. P. M., Garibaldi, L. A., Potts, S. G., and van der Putten, W. H. (2019) 'Ecological Intensification: Bridging the Gap between Science and Practice', *Trends in Ecological Evolution*, vol. 34(2), pp. 154–166.

Li, H., Guo, H., Helbig, M., Dai, S., Zhang, M., Zhao, M., Peng, C., Ming, X. and Zhao, B. (2019) 'Does Direct-seeded Rice Decrease Ecosystem-scale Methane Emissions?—A Case Study from a Rice Paddy in Southeast China', *Agricultural and Forest Meteorology*, vol. 272–273, pp. 118–127.

Leahy, S., Clark, H. and Reisinger, A. (2020) 'Challenges and Prospects for Agricultural Greenhouse Gas Mitigation Pathways Consistent with the Paris Agreement'. Available at: https://www.frontiersin.org/articles/10.3389/fsufs.2020.00069/full

Loboguerrero, A. M., Campbell, B., Cooper, P., Hansen, J.W., Rosenstock, T. and Wollenberg, E. (2019) 'Food and Earth Systems: Priorities for Climate Change Adaptation and Mitigation for Agriculture and Food Systems', *Sustainability*, vol. 11, p. 1372.

Lowder, S. K., Skoet, J. and Raney, T. (2016) 'The Number, Size, and Distribution of Farms, Smallholder Farms, and Family Farms Worldwide', *World Development*, vol. 87, pp. 16–29.

Lynch, J. and Garnett, T. (2021) 'Policy to Reduce Greenhouse Gas Emissions': Is Agricultural Methane a Special Case? Available at: https://onlinelibrary.wiley.com/doi/10.1111/1746-692X.12317?af=R

Leahy, S., Clark, H. and Reisinger, A. (2020) 'Challenges and Prospects for Agricultural Greenhouse Gas Mitigation Pathways Consistent with the Paris Agreement', *Frontiers of Sustainable Food Systems*, vol. 4, pp. 69.

Martin, E. A., Dainese, M., Clough, Y., Báldi, A., Bommarco, R., Gagic, V. and Steffan-Dewenter, I. (2019) 'The Interplay of Landscape Composition and Configuration: New Pathways to Manage Functional Biodiversity and Agroecosystem Services Across Europe', *Ecology Letters*, vol. 22 (7), pp. 1083–1094.

McKinsey & Company (2020) 'Agriculture and Climate Change – Reducing Emissions through Improved Farming Practices'. Available at: https://www.mckinsey.com/~/media/mckinsey/industries/agriculture/our%20insights/reducing%20agriculture%20emissions%20through%20improved%20farming%20practices/agriculture-and-climate-change.pdf

MacLeod, M., Eory, V., Gruère, G. and Lankoski, J. (2015) 'Cost-Effectiveness of Greenhouse Gas Mitigation Measures for Agriculture: A Literature Review', OECD Food, Agriculture and Fisheries Papers, No. 89, OECD Publishing, Paris. Available at: http://dx.doi.org/10.1787/5jrvvkq900vj-en

Mwangi, M., Kituyi, E., Ouma, G. and Macharia, D. (2020) 'Indicator Approach to Assessing Climate Change Vulnerability of Communities in Kenya: A Case Study of Kitui County', *American Journal of Climate Change*, vol. 9, pp. 53–67.

Nagothu, U. S. (2015) The Future of Food Security: Summary and Recommendations. In: Nagothu, U. S. (ed.) *Food Security and Development: Country Case Studies*. London: Routledge, pp. 274.

Nagothu, U. S., Bloem, E. and Andrew, B. (2018) Agricultural Development and Sustainable Intensification: Technology and Policy Innovations, In: Nagothu, U. S (ed.) *Agricultural Development and Sustainable Intensification, Technology and Policy Challenges in the Face of Climate Change*, London: Routledge, pp. 1–22.

Nicastro, R. and Carillo, P. (2021) 'Food Loss and Waste Prevention Strategies from Farm to Fork', *Sustainability*, 13, pp. 5443.

NORR (2021) 'Climate Change: Atmospheric Carbon Dioxide'. Available at: https://www.climate.gov/news-features/understanding-climate/climate-change-atmospheric-carbon-dioxide

OECD (2021) 'Enhancing Climate Change Mitigation Through Agriculture'. Available at: https://www.oecd.org/publications/enhancing-the-mitigation-of-climate-change-though-agriculture-e9a79226-en.htm

Raphaela, J. P. A., Calonegoa, J. C., Milorib, D. M. B. P. and Rosolema, C. A. (2016) 'Soil Organic Matter in Crop Rotations Under No-till', *Soil and Tillage Research*, vol. 155, pp. 45–53.

Rapsomanikis, G. (2015) 'The Economic Lives of Smallholder Farmers. An Analysis Based on Household Data from Nine Countries'. Available online at: http://www. fao.org/3/a-i5251e.pdf

Reardon, T., Echeverria, R., Berdegué, J., Minten, B., Liverpool-Tasie, S., Tschirley, D. and Zilberman, D. (2019) 'Rapid Transformation of Food Systems in Developing Regions: Highlighting the Role of Agricultural Research & Innovations', *Agricultural Systems*, vol. 172, pp. 47–59.

Reid, S., Hayes, J. P. and Stibbe, D. T. (2014) 'Platforms for Partnership: Emerging Good Practice to Systematically Engage Business as Partner in Development'. Available at: https://thepartneringinitiative.org/publications/research-papers/platforms-for-partnership-emerging-good-practice-to-systematically-engage-business-as-a-partner-in-development/

Reay, D. S., Davidson, E. A., Smith, K. A., Smith, P., Melillo, J. M., Dentener, F. and Crutzen, P. J. (2012) 'Global Agriculture and Nitrous Oxide Emissions', *Nature Climate Change*, vol. 2(6), pp. 410–416.

Reisinger, A., Clark, H., Abercrombie, R. M., Aspin, M., Ettema, P., Harris, M. and Sneath, G. (2018) *Future Options to Reduce Biological GHG Emissions On-farm: Critical Assumptions and National-scale Impact*. Report prepared for the Biological Emissions Reference Group. Wellington: Ministry for Primary Industries, 80'. Available at: https://www.mpi.govt.nz/dmsdocument/32128/send

Richards, M. (2019) 'National Plans to Address Adaptation and Mitigation in Agriculture: An Analysis of Nationally Determined Contributions. CCAFS Dataset', Wageningen: CGIAR Research Program on Climate Change, Agriculture and Food Security (CCAFS). Available at: https://hdl.handle.net/10568/101189

Roos, E., Bajzelj, B., Smith, P., Patel, M., Little, D. and Garnett, T. (2017) 'Greedy or Needy? Land Use and Climate Impacts of Food in 2050 Under Different Livestock Systems', *Global Environmental Change*, vol. 47, pp. 1–12.

Smith, P. (2016) 'Soil Carbon Sequestration and Biochar as Negative Emission Technologies', *Global Change Biology*, vol. 22/3, pp. 1315–1324.

Tziva, M., Negro, S. O., Kalfagianni, A. and Hekkert, M. P. (2020) 'Understanding the Protein', Transition: The Rise of Plant-based Meat Substitute'. Available at: https://www.sciencedirect.com/science/article/pii/S2210422419302552

United Nations (2020) 'Sustainable Development Goals'. Available at: https://www.un.org/sustainabledevelopment/climate-change/

UNEP (2020) 'Emissions Gap Report'. Available at: https://wedocs.unep.org/bitstream/handle/20.500.11822/34438/EGR20ESE.pdf

UNFCCC (2015) 'Adoption of the Paris Agreement', Available at: https://unfccc.int/resource/docs/2015/cop21/eng/l09.pdf

UNFCCC (2018) 'Climate Action Now Summary for Policymakers 2018'. Available at: https://unfccc.int/resource/climateaction2020/spm/introduction/index.html

UNFCCC (2021) 'What is the Kyoto Protocol'. Available at: https://unfccc.int/kyoto_protocol

Vermeulen, S. J., Aggarwal, P. K., Ainslie, A., Angelone, C., Campbell, B. M. and Challinor, A. J. (2012) 'Options for Support to Agriculture and Food Security Under Climate Change'. *Environmental Science and Policy*, vol. 15, pp. 136–144.

Vermeulen, S. J., and Wollenberg, E. K. (2017) 'A Rough Estimate of the Proportion of Global Emissions from Agriculture Due to Smallholders. CCAFS Info Note'. Available at: https://ccafs.cgiar.org/publications/rough-estimateproportion-global-emissions-agriculture-due-smallholders#.XfFhLOgzaUk

Win, E. P. and Win, K. K. (2020) 'GHG Emissions, Grain Yield and Water Productivity'. Available at: https://public.wmo.int/en/media/press-release/weather-related-disasters-increase-over-past-50-years-causing-more-damage-fewer

WMO (World Meteorological Organization) (2021) 'Greenhouse Gas Bulletin: Another Year, Another Record'. Available at: https://public.wmo.int/en/media/press-release/greenhouse-gas-bulletin-another-year-another-record

Wollenberg, E. (2017) 'The Mitigation Pillar of Climate-Smart Agriculture (CSA): Targets and Options', *Agricultural Development*, vol. 30, pp.19–22.

WSJ (2021) 'COP26: What Is the Climate Summit in Glasgow and Why Does It Matter?' Available at: https://www.wsj.com/articles/cop26-glasgow-climate-summit-2021-11611254971

2 Precision-based soil and nutrient management tools for enhancing soil health while reducing environmental footprint

Amaresh Kumar Nayak, Sangita Mohanty, Mehreteab Tesfai, Rahul Tripathi, Anjani Kumar, and Udaya Sekhar Nagothu

Introduction

Soil is one of our most important natural resources that provide us with vital goods and services to sustain life on land. If soils are healthy and sustainably managed, they can provide adequate food, clean water, habitats for biodiversity, and other important ecosystem services while contributing to climate resilience, adaptation, and mitigation (Stolte et al., 2016). Soils act as source and sink for greenhouse gases (GHGs) such as carbon dioxide, methane, and nitrous oxide (Oertel et al., 2016). As a source, soil emits nitrous oxides from applied nitrogen fertilizers and as sink, soil increases carbon sequestration including carbon storage via fixation and organic fertilizer addition while reducing environmental footprints from rice cultivation.

Rice is one of the staple foods of India that occupies about 24% of its total cropped area and contributes 42% of total food grain production. Being input-intensive crop, rice requires around 15–20 kg of nitrogen to produce 1,000 kg of grain (e.g., Peng et al., 2010). Most Indian soils contain low-to-medium plant available nitrogen (N), and therefore the yield potential of rice or other crops largely depends on the exogenous application of nitrogen fertilizers (Panda et al., 2019). Rice cultivation alone accounts for 37% of the total N fertilizer consumption in India (FAI, 2018). However, more than 60% of this applied N is lost to environment in the form of N_2O, NH_3, and NO_3. The conventional practice of rice cultivation in India involves ponding water between 5 and 7 cm depth in the soil for a considerable part of the growing period. Such soil micro-environment accelerates the processes of nitrogen transformation and its losses through nitrification, denitrification, volatilization, leaching, and runoff, which has resulted in low nitrogen use efficiency (NUE). NUE of a cropping system is defined as 'the proportion of all N inputs that are removed in harvested crop biomass, contained in recycled crop residues, and incorporated into soil organic matter and inorganic N pools' (Cassman et al., 2002). As we are aware, use of fertilizer N for crop

DOI: 10.4324/9781003273172-2

production influences soil health primarily through changes in organic matter content, microbial life, and acidity in the soil (Bijay Singh, 2018).

Soil health and nitrogen management

Soil health is defined as the continued capacity of soils to function properly and pro-vide the required ecosystem services and goods (EC, 2021). We take the soil services for granted, but in fact soils are non-renewable and a threatened resource globally. The effects of climate change are putting further pressure on the soil resources and overall health of the soil. Most of the agricultural soils in the world are unhealthy, mainly because of unsustainable soil management practices. Some of the soil health improving practices generally adopted by farmers include conservation agriculture (*minimum soil tillage, residue mulching, crop rotations*), *intercropping with legumes, trees alley cropping, green manuring,* and *composting,* which have a potential to reduce the exogenous application of mineral N fertilizer applications (refer Chapter 8 of this book). However, farmers do not receive any incentives for adopting such soil practices.

Nitrogen is one of the most limiting nutrients for rice production and is a sig-nificant source of N_2O (a potent greenhouse gas: GHG) from agricultural soils. The direct N_2O emission from synthetic N fertilizers used in agriculture has been estimated to be 0.9 N_2O Tg N per year (Syakila and Kroize, 2011). One of the non-point sources of pollution that cause serious threat to water environments is reactive N losses in the form of NO_3.

During the green revolution, nitrogen fertilizers were one of the key drivers to food production and have transformed the Indian agriculture to become the worlds' second leading food grain producer. Nitrogen fertilizer consumption in Indian agriculture witnessed a whooping increase from 3.4 million tonnes (in the late 1970s) to 16.9 million tonnes in 2017–2018 (Figure 2.1). The trend will con-tinue to increase further to attain the national yield target of 350 million tonnes

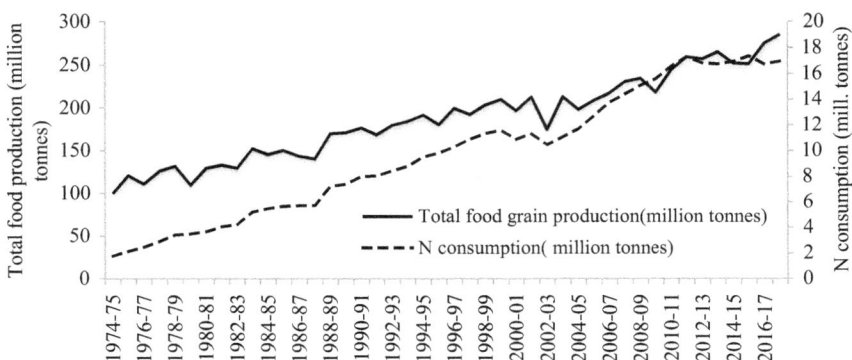

Figure 2.1 Trend of total food grain production and fertilizer N consumption in India: 1974–2016.

Source: Fertilizer statistics (2018–2019). Data adapted from https://www.faidelhi.org/statistics/statistical-database

of food grains (cereals and pulses) in India by 2050, to feed around 1.8 billion people (Kumar and Sharma, 2020). During these years, not only the consumption of fertilizer N will increase tremendously but also the flow of reactive N from agroecosystems and associated environmental losses.

However, the increased food production during the Green Revolution (e.g., by applying nitrogen fertilizers) was at the expense of environmental degradation. It has caused severe degradation of land and water resources, soil pollution, and high levels of GHG emissions (Rahman, 2015). To address these challenges, research on nitrogen fertilizer management focused more on nutrient stewardship principles that entail *right rate, right time, right methods, and right source (4R)* of nutrient application (IFA, 2009).

The above-mentioned 4R principles of nutrient stewardship (Box 2.1) are not new. This chapter, however, attempted to present and discuss context-specific precision nutrient management tools and techniques that can effectively support the 4R principles in the context of rice cultivation to improve soil health while reducing environmental footprints. The 4R principles could be applied using precision-based tools such as remote sensing, Geographic Information System (GIS), Global positions systems (GPS), and simulation modelling. One of the key factors in implementing precision nutrient management is the ability to provide timely information regarding spatial distribution of crop N status within a field. From this perspective, determining plant N concentration by proximal or remote sensing techniques is much more appealing than the traditional destructive chemical analyses on soil/plant samples considering the cost and time required.

Box 2.1   4R Nutrient Stewardship principle

4R Nutrient Stewardship is a new innovative approach for fertilizer best management practices that considers economic, social, and environmental dimensions of fertilizer management and is essential to sustainability of agricultural systems. The concept is simple: *apply the Right source of nutrient, at the Right rate, at the Right time, and in the Right place.* All farmers – irrespective of the size of farm, knowledge and awareness levels – consider what fertilizer to apply, how much, when, and how before making a fertilizer application decision in any crop. The 4R Nutrient Stewardship principles connect these fertilizer application decisions to scientific principles and guide the application decisions to specific crops, soils, and local site. Moreover, the four 'rights' provide a simple checklist to assess whether a given crop has been fertilized properly. Asking 'was the crop given the right source of nutrients at the right rate, time, and place?' helps farmers and advisors to identify opportunities for improvement in fertilizing each specific crop in each specific field.

Source: Adapted from Sapkota et al. (2016)

Moreover, a declining trend of NUE in terms of partial factor productivity and recovery efficiency of nitrogen in the Indian agriculture (Bijay Singh, 2017) calls for a more precise nutrient stewardship approach in the context of rice cultivation that ensures environmental sustainability and improves soil health. N fertilizer recovery efficiency (RE_N) is defined as 'percentage of fertilizer-N recovered in aboveground plant biomass during the growing season' and partial factor productivity of N (PFP_N) also called agronomic N use efficiency (AE_N) is 'the ratio of crop yield per unit of applied N fertilizer' (Cassman et al., 2002). Understanding the reasons for the declining trends of NUE and the prognosis for improving them depends on knowledge of the factors that govern N demand and supply in rice cropping systems.

The objectives of the chapter are to review and analyse research findings in precision nutrient management tools and techniques with special focus to N fertilizer use efficiency in rice cultivation. The chapter also discusses soil management interventions that can improve soil health while at the same time reduce environmental footprints and their policy implications for large-scale adoption.

Methodological framework

The 4R Nutrient Stewardship framework was applied in the Norwegian funded Resilience project (www.resilienceindia.org) in the case studies of Odisha state (India) by adopting best management practices for soil and nutrients focusing on Nitrogen. The 4R principles aim to enhance production, increase NUE, and increase farmer profitability while reducing environmental footprints (Figure 2.2).

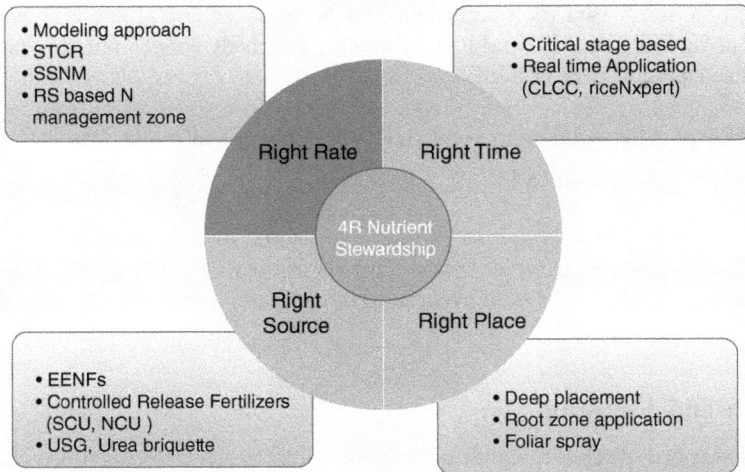

Figure 2.2 Framework of 4R Nutrient Stewardship for enhancing N use efficiency.
Source: Authors' own compilations. Soil Test-based Crop Response (STCR), Site-Specific Nutrient Management (SSNM), Remote Sensing (RS), Customized Leaf Colour Chart (CLCC), Enhanced Efficiency of N Fertilizers (EENFs) that includes Sulphur-Coated Urea (SCU), Neem-Coated Urea (NCU), and Urea Super Granules (USG).

Table 2.1 Basic principles of precision-based tools/techniques for N management

Precision-based tools	Principles and applications
Soil test-based crop response approach	Precise fertilizer recommendations are made after the establishment of a significant relationship between soil test values, added fertilizer nutrients, and crop response for a particular soil type (Singh et al., 2021)
Site-specific nutrient management approach	Supplying plants with nutrients to optimally match their inherent spatial and temporal needs for supplemental nutrients by using different tools such as remote sensing, GPS, and GIS systems (Verma et al., 2020)
Leaf colour chart	A diagnostic tool used to determine N level in rice plants relative to greenness of the leaves, containing at least four panels of colour, ranging from yellowish green to dark green (Nayak et al., 2013)
Sulphur-coated urea	Slow-release fertilizer made by coating urea with sulphur and wax that increases NUE, improves plant growth, and reduces water pollution (Shivay et al., 2016)
Neem-coated urea	Nitrification inhibitor which increases yield, uptake, and use efficiency of applied N fertilizer in rice (Meena et al., 2018)
Urea super granules	Fertilizer applied at 8–10 cm soil depth saves 30% N, increases absorption rate, improves soil health, and ultimately increases rice yield (Sarker et al., 2012)

Source: Authors' own compilation from different literatures.

The precision-based tools/techniques for N management (Table 2.1) were assessed using PFP_N or AE_N and RE_N which are relatively easy to measure and are often used as indicators of NUE.

$$PFP_N \ or \ AE_N = \text{kg grain yield (ha}^{-1} / \text{kg}^{-1}\text{N applied ha}^{-1}) \tag{2.1}$$

$$RE_N = (U_N - U_0)/F_N) \times 100 \tag{2.2}$$

where U_N is the plant N uptake (kg ha^{-1}) measured in aboveground biomass at physiological maturity in a plot that received N at the rate of FN (kg ha^{-1}), and U_0 is the N uptake measured in aboveground biomass in a plot without the addition of fertilizer N.

Results and discussion

In this section, selected research findings on NUE in rice growing areas of India and other Asian countries were reviewed and discussed. In addition, research data from the Resilience project (www.resilienceindia.org) were analysed to show the performance of precision-based tools and soil management practices in Odisha. The soil management practices used were conservation agriculture (e.g., minimum tillage, no-till farming), crop diversification (including crop rotations and intercropping), crop residues mulching, composting, green manuring, and biofertilizers.

Precision tools for nutrient management

Several precision tools and techniques were tested in the Resilience project to assess and guide farmers in the real-time crop N need in line with the 4R Nutrient Stewardship principles. These include from simple/easy-to-use and inexpensive diagnostic tools like the leaf colour chart to highly sophisticated optical sensors for precision-based fertilizer recommendations using drone-mounted sensors.

i Leaf colour chart

The leaf colour chart (LCC) monitors relative greenness of a rice leaf and can be used as an indicator of the crop N status for determination of crop N demand in the season. However, the critical colour for N application may vary with cultivars, agro-climatic situations, and growing season and varieties grown. This requires cultivar specific standardization of LCC and calibration and validation of critical colour code. In response to this, a customized leaf colour chart (CLCC) was developed by the Indian Council of Agricultural Research, National Rice Research Institute (ICAR-NRRI) based on leaf colour analysis of hundreds of rice varieties (Nayak et al., 2013). Results from field trials showed yield advantages of 0.5–0.7 t ha^{-1} with CLCC-based N application (25% lower than the normal recommended dose of N) (Nayak et al., 2017; Nayak et al., 2018), which implies less cost of fertilizers for farmers and reduced environmental footprint.

The CLCC-based urea-N application increased yield by 10–13% in direct seeded rice (DSR) and 10–11% in puddled transplanted rice (PTR) over conventional systems (Mohanty et al., 2017). However, when neem-coated urea (NCU) was applied based on CLCC recommendations, the yield advantage further maximized to 21–23% for DSR and 15–16% for PTR (Mohanty et al., 2021). A possible explanation for higher yield could be attributed to the synchronized application of NCU, which reduces nitrification but increases plant available N, which results in higher PFP_N or AE_N. The performance of LCC-based N application compared to blanket application in terms of increasing grain yield and enhancement of NUE is shown in Table 2.2.

Depending upon crop management and agro-climatic conditions, the real-time N application using LCC can potentially increase yield by 10–22% (Mohanty et al., 2018) and save N fertilizer consumption by up to 50 kg ha^{-1} (Bijay Singh et al., 2003). Similar studies by Kumar et al. (2018) also observed significantly higher grain yield of rice with CLCC-based N application, compared to blanket method of N application.

The values for NUE ranged from 7.9% (RE_N), 44% (AE_N), to 57% (PPF_N) under no basal/basal + LCC (3–4) based top dressing (Table 2.2). The low RE could be a consequence of greater N losses via nitrate leaching and N_2O emission. This is in line with results from a study by Andrés-Barbieri et al. (2018) who reported that RE was 29% (on average) for urea which is low because of greater N mineralization from organic matter, N losses by denitrification, or immobilization.

An application of N fertilizer with NCU using CLCC readings reduced N_2O emission by 13–14% in DSR and 16–23% in PTR (Mohanty et al., 2017).

Table 2.2 Performance of LCC-based N application compared to blanket N application

Application method	Grain yield increase (%)	NUE enhancement (%)	References
Basal + LCC (4.0) based top dressing	20	15–57 (PPF$_N$)	Ali et al. (2017)
Basal + LCC (4.0) based top dressing	22	7.9 (RE$_N$)	Mohanty et al. (2018)
Basal + CLCC (≤ 3.0) based top dressing	10–13	8.9–12.4 (RE$_N$)	Mohanty et al. (2021)
No basal + LCC (4.0) based top dressing	12–16	19–44 (AE$_N$)	Shukla et al. (2004)
No basal + LCC (≤ 4.0) top dressing	11.3	13 (RE$_N$)	Bhatia et al. (2012)

RE$_N$: recovery efficiency of N, AE$_N$: agronomic efficiency of N, PPF$_N$: partial productivity factor of N.
Source: Adapted from several sources.

Though DSR reduces CH_4 emission as it uses less water during initial cropping, it can increase N_2O emissions. The CLCC-based NCU application addresses the trade-offs between CH_4 and N_2O emissions by translating their emissions into global warming potential (GWP). In other words, minimum cumulative radiative forcing of the two gases on GWP is the possible option to minimize the trade-off and effect of GHG emissions (Susilawati et al., 2019). Moreover, the CLCC-based NCU application minimized yield loss, and resulted in 20–25% reduction of GHG index through DSR (Mohanty et al., 2017). Thus, the GHG index (GHGI) that compares GWP and grain yield shows a lower value for DSR, implying that DSR mitigates GHG emission and produces more rice.

Research on LCC-based N application in other rice growing areas of Asia also showed similar results about saving of N. For instance, savings of N by up to 25% by Alam et al. (2006) and 8.3% N by Sen et al. (2011) using LCC-based N management in different rice genotypes compared to the prescribed dose of N application (120 kg N ha^{-1}). Moreover, the rice grain yields were 4.8 t ha^{-1} and 4.3 t ha^{-1} with variety *NDR-359* and *Sarju-52*, respectively.

Since the CLCC is user friendly to small-scale farmers, several state government departments, Indian Council of Agricultural Research institutes, and state agricultural universities in India have taken the initiatives to develop region-specific CLCC and are upscaling it through various schemes (Figure 2.3). Today, more than 500,000 units (cards) of the Indian Rice Research Institute (IRRI)-LCC type have been produced and distributed to farmers through collaboration with the National Agricultural Research and Extension Systems. The IRRI-Hyderabad, India centre developed a five-panel modified LCC for irrigated rice. Similarly, the Punjab Agricultural University (PAU) has come up with a six-panel LCC for major food crops such as rice, wheat, and maize. Studies in the Punjab state of India indicated an average saving of 30 kg N ha^{-1} due to the use of the PAU-LCC.

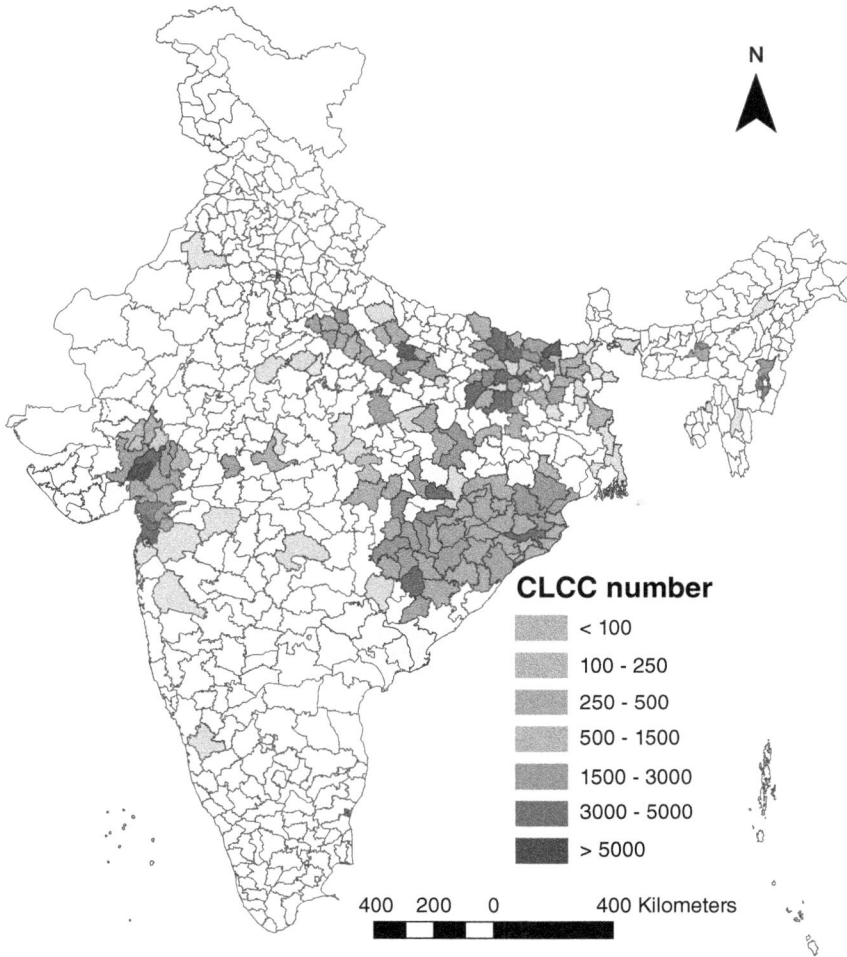

Figure 2.3 Spread of CLCC across India. Source: Authors' own compilation.

The CLCC developed by ICAR-NRRI (Cuttack, Odisha) provides cultivar-specific N recommendation for rainfed lowland, submerged/flood prone lowland, rainfed upland, and irrigated rice of eastern India (Nayak et al., 2013). So far, more than 200,000 units of CLCCs were distributed in several states of eastern India.

ii   Android-based mobile app: riceNxpert
     Android-based real-time N application 'riceNxpert' app developed by the ICAR-NRRI under the Resilience project gives recommendations for in-season N application on the basis of leaf colour analysis. The riceNxpert is a cost-effective app for real-time N application, is user friendly, has a potential for

wider adaptation, and avoids the need for purchase CLCC every year. Research results showed that riceNxpert-based N application enhanced rice yield by 6. 6–23.8% over RDN application and by 38% over normal farmers' practice (Nayak et al., 2021). This implies a yield advantage of 0.5–1.1 t ha^{-1} over RDN, and the riceNxpert-based N application in fact can give monetary benefits ranging from INR 9,000 ha^{-1} to INR 20,000 ha^{-1} (~US\$118–263) over RDN. Moreover, riceNxpert has a potential to cut down N use by 15–25% in rice production and hence bring significant reduction of India's subsidy bill for urea.

Farmer-led field experiments using riceNxpert in eastern India have increased crop yield and PPF$_N$ by 12–13% compared to RDN (Figure 2.4). However, trials conducted using riceNxpert-based N application in the research farm have enhanced PPF$_N$ by 30–40% and RE$_N$ by 9–15% compared to RDN. The performance between riceNxpert and CLCC on crop yield and NUE (AE$_N$ or PFP$_N$) was compared, and results are presented in Figure 2.4.

iii   Chlorophyll meter and SPAD meter

The chlorophyll meter and the soil plant analysis development (SPAD) meter are also real-time N application tools/techniques. The chlorophyll meter provides the relationship of chlorophyll content of leaf with that of N contents which is more of quantitative indicator for crop N status. The SPAD meter estimates chlorophyll content of leaf in the field by measuring difference in light attenuation at 430 (spectral transmittance peaks for chlorophyll a, b) and 750 nm wavelength, near-infrared spectral region (Schröder et al., 2000). SPAD with threshold 36 could save N fertilizer by 20–35% while maintaining the same level of yield and improves RE$_N$ by 14–19% (Ghosh et al., 2013).

iv   Comparative performance of LCC, riceNxpert, and SPAD

The performances of real-time N application tools such as riceNxpert, CLCC, and SPAD meter were compared under field trials in the Resilience project (Figure 2.5). The difference in the performances of the three tools in

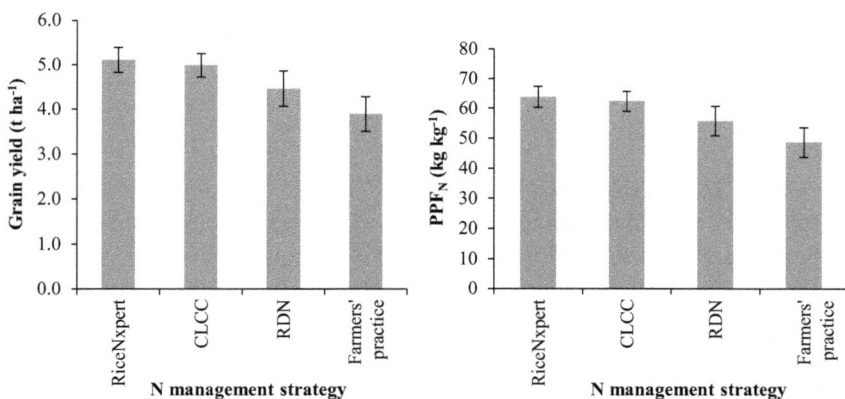

Figure 2.4 Grain yield and PFP$_N$ of rice under different N application strategies.
Source: Field data from Resilience project.

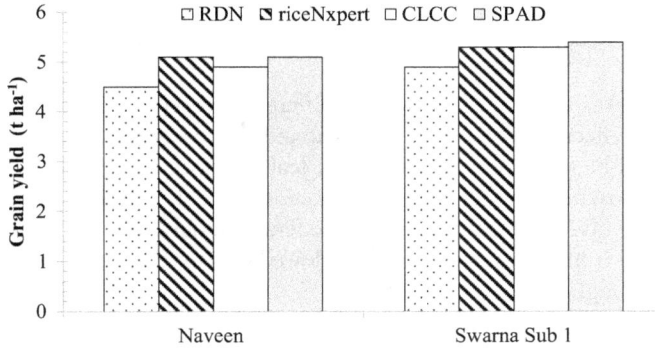

Figure 2.5 Comparative performance of real-time N application tools in relation to grain yield of rice varieties: *Naveen* and *Swarna Sub 1*.
Source: Field data from Resilience project.

terms of grain yield in the two rice varieties (*Naveen* and *Swarna Sub*) was non-significant (Figure 2.5). However, the mean difference in grain yield was statistically significant (*P<0.05*) between the tools and RDN. There was on average an 8–13% increase in grain yield using the precision-based tools over RDN. The grain yield of *Swarna Sub* variety was higher than *Naveen variety* in all real-time N application tools including RDN but not significant (*P<0.05*).

Studies comparing real-time N management tools between SPAD and CLCC are limited and showed varying results. For instance, Patil et al. (2018) observed that N application using LCC (threshold 4) produced 8.5% higher yields and using SPAD (threshold 40) produced 5% higher yields than RDN application. An LCC-based application saved 30 kg N ha^{-1}, while SPAD saved only 10 kg N ha^{-1} compared to RDN application. This indicates the need for situation-specific standardization of SPAD threshold for N application.

An application of 30 kg N ha^{-1} at ≤3.0 LCC led to additional application of 42 kg N ha^{-1} over RDN and contributed to a yield increase by 3.5–19.8%, whereas an application of 30 kg N ha^{-1} at ≤35 SPAD produced 0.6–15.4% higher yield than RDN (Jahan et al., 2018). This implies that rice crop showing less than 4 LCC reading requires more N application than using SPAD meter. Other studies conducted by Ali et al. (2015) showed that applying 30 kg N ha^{-1} at <4 LCC saved up to 40 kg N ha^{-1}, whereas using SPAD (reading < 37) savings increased to 70 kg N ha^{-1} while producing similar yield. In other words, applications of N using SPAD (with 35–37) save more N than LCC (<4) based applications without yield loss.

Other precision tools and approaches

Green Seeker is another precision tool used to assess the real-time crop N (Box 2.2). It is an optical sensor-based tool that can be used to calculate season N requirement as per site-specific need of the crop.

Box 2.2 Green Seeker: Principles and Applications

Green Seeker measures normalized difference vegetation index (NDVI) based on reflectance of radiation in the red and near-infrared bands. The NDVI can be used to predict biomass, leaf N concentration, plant N uptake, photosynthetic efficiency, and potential yield. The N management strategy involves the application of a moderate amount of N at planting and crown root initiation stages. This is followed by a corrective Green Seeker-guided N application at weeks 5–6 or 7–8 which resulted in higher yield and NUE than the blanket recommendation. The adoption of sensor-based N management strategy, consisting of 30 kg N ha^{-1} at transplanting and 45 kg N ha^{-1} at active tillering stage followed by a Green Seeker-guided dose at panicle initiation stage, rendered 5–22% higher N recovery efficiency.

Source: Bijay-Singh et al. (2015)

There are also models such as Quantitative Evaluation of Tropical Soil Fertility (QUEFTS), decision support tools (remote sensing, GIS), enhanced efficiency of N fertilizers (e.g., SCU, NCU, and USG), and precision-based approaches such as STCR and SSNM, which are coming into use these days to assess the real-time crop N need (refer Table 2.1). Recommendations from the precision tools must be put into wider practice in the cultivation of major cereal crops such as rice and wheat to increase environmental sustainability and reduce GHG emissions.

Site-specific nutrient management (SSNM)

Different models such as QUEFTS and tools (e.g., remote sensing, GPS) are used to implement the SSNM approach as shown in Table 2.3. The major steps in SSNM are fixing an attainable yield target and estimating nutrients required to attain the yield by (i) harnessing the available nutrient supply from all possible sources, and (ii) filling the gap between crop nutrient need and indigenous nutrient supply (Verma et al., 2020).

i   Models and decision support tools
Several dynamic crop growth simulation models have been developed in recent years to predict the crop yield which is crucial for estimating the nutrient needs. The QUEFTS model has been effectively applied to a variety of crops, including maize and rice. The algorithms of SSNM generated based upon QUEFTS model are embedded into a web-based decision support tool known as nutrient manager for rice (Buresh et al., 2014), which was later renamed as Rice Crop Manager (RCM). The RCM can provide field-specific recommendations with respect to rate and time of fertilizer application.
   Web-based decision support tools such as Nutrient Expert and RCM have a potential to scale up precise nutrient recommendations to a wider area.

Table 2.3 Examples of previous research findings using SSNM approaches

SSNM approaches/tools	*Key research finding*
Models (e.g., QUEFTS) and decision support tools	Field evaluation ($n = 209$) in six agro-climatic zones of Odisha showed RCM recommendation provided yield advantage of 23% over farmer's practice and income increased by US$188 ha^{-1} per season on average (Sharma et al., 2019)
Soil test-based crop response (STCR)	STCR enhanced grain yield by 20% and increased RE$_N$ by 18% under DSR (Singh et al., 2021)
Remote sensing and GIS-based N application	Multispectral data provided by the MODIS satellite were used to predict leaf N status in rice. It indicated the in-season N need of the crop which varied between 60 and 120 kg N ha^{-1} for rice growing areas in Odisha. This enhanced rice yield by 8–12% over blanket recommendation (Tripathi et al., 2017)

Promising initiatives are being undertaken by the NARES and state agricultural extension departments to disseminate RCM-based recommendations through public extension services, private initiatives such as the village knowledge centres (promoted by Resilience project), and e-service facilities. These extension services are available at village level in India. More than 175,000 RCM-based nutrient recommendations have been generated to farmers in Odisha that has increased the grain yield by 0.3–0.8 t ha^{-1} compared to the normal farmers' fertilizer practices and by 0.2–0.4 t ha^{-1} over blanket fertilizer recommendation (IRRI, 2019).

ii Soil test-based crop response

It is evident that there is a linear relationship between crop yield and nutrient uptake until a certain threshold. This linear relationship is used to estimate the fertilizer requirement of a crop considering the efficiency of soil and fertilizer nutrients. Based on the targeted yield approach, the ICAR initiated All India Coordinated Research Project Soil Test Crop Response (AICRP-STCR). The objective is to develop fertilizer prescription equations for different crops under different agroecological regions. STCR studies have developed relationship between crop yield, on the one hand, and soil test estimates and fertilizer inputs, on the other hand, using 'targeted yield approach' (Jat et al., 2015). An STCR-based approach of nutrient application has advantage in terms of increasing nutrient response ratio over RDN application and farmer's practice. Despite this, soil and plant leaf tests are time-consuming, inconvenient, and expensive, and do not account spatial/temporal variability nutrient contents within a field. Thus, remotely controlled drone-mounted sensors and associated tools are required to implement precision-based N application over space and time.

iii Drone-mounted sensors, remote sensing, and GIS-based N application

Soil and plant sampling and tests using sensors attached on drones can collect remotely sensed data from plant canopies, which can be used to measure canopy greenness and precise N topdressing recommendations (Huuskonen and Oksanen, 2018). Drone-mounted sensors, geospatial sciences, GIS, GPS, big data analytics, and ICTs have a distinct advantage in terms of capturing the variability of agricultural systems for spatio-temporal monitoring of crop and soil.

Remote sensing methods have been proposed to help in precision farming to gather data, and with proper analytics, the growth during the season can be monitored. Remote sensing provides spatially dense information that can be used to assess the crop N status and understand and predict spatially variable crop N needs. Tripathi et al. (2017) established a relationship between NDVI (obtained from MODIS image) and leaf N concentration (field measurements) using univariate regression analysis. They developed N fertilizer recommendation map using MODIS NDVI and leaf area index for rice growing areas in Odisha.

The precision-based tools/techniques used for N management could promote the effective implementation of soil management practices thereby improving soil health and reducing environmental footprints in rice cropping systems.

Agroecological-based approaches for improving soil health

Selected conservation agriculture (CA) and/or agroecological (AE) based and integrated nutrient management practices that have the potential to improve soil health were reviewed and discussed below.

Conservation agriculture and/or agroecological practices

The three major principles of CA are (i) *minimal soil disturbance (minimum/zero tillage)* by direct sowing into the soil cover without seedbed preparation (refer Chapter 6), (ii) *keeping soil permanently covered by residue retention or growing cover crops*, and (iii) *diversification including crop rotations and intercropping*.

i   *Zero tillage and/or minimum tillage:* There are several research studies that demonstrate minimal soil disturbance improves soil health and reduces carbon or nitrogen losses. For example, a study conducted by Lal et al. (2019) showed that zero tillage (ZT) lowered energy use by 56%, carbon footprints by 39%, and N_2O emissions by 20% compared to conventional tillage. Zero tillage also contributes to improving soil health by enhancing the major plant nutrients including N, labile pool of carbon and enzymatic activities in soil. A study by Dash et al. (2017) found that the change in soil organic carbon (SOC) stock was significantly higher under ZT transplanted rice by 5–7% in the dry season and 8–10% in the wet season, compared to conventional tillage practices. About 12% less methane emission was recorded under ZT treatments over control. In the long run, ZT can aid in enhancing soil health and reducing the environmental footprint of a rice cropping system despite a 10–15% yield penalty in early years of the experiment. However, studies also (e.g., Yadav et al., 2021) reported that conventional tillage disrupts soil structure and accelerates soil carbon loss by exposing the inter/intra-aggregate spaces of carbon to aerobic microbes for rapid oxidation (Al-Kaisi and Yin, 2005).

ii *Crop diversification (including crop rotations/intercropping)*: The rice-based monocropping systems in India and other rice growing regions could benefit by growing diverse crops/varieties over time and/or space. These benefits include increased productivity, increased profitability, and reduced climate change risks. Farmer-led demonstration trials were conducted in Odisha under the Resilience project. The trials showed that the diversification of rice fallow with toria (*Brassica campestris* L.) and green/black gram rendered additional yield of 2 t ha^{-1} and increased system productivity from 4 t ha^{-1} to 6 t ha^{-1}. Other studies, e.g., growing dry season crops such as green gram and black gram, had reduced energy use and carbon footprints (Lal et al., 2020). An inclusion of black gram and green gram in a rice-based cropping system enhanced yield of subsequent crop and resulted in 0.04% increase in SOC content (Porpavai et al., 2011) and reduced GHG emission (Dash, 2019).

iii *Crop residue mulching*: Crop residue mulch-based ZT system can save energy in the form of fuel and labour, reduce carbon footprints, and improve net farm income, soil health, and environmental quality. However, organic supplements are scarce in dry areas of the tropics and farmers use the crop residues for feed, fuel, and/or construction materials. Burning of crop residues emits GHGs, and causes problems of air pollution, health hazards, and loss of nutrients which is a matter of serious concern in countries like India. Crop residue retention in the form of residue mulching using rice straw enhanced crop production, microbial biota, and enzymatic activities of soil (Lal et al., 2019). It also has a potential to reduce carbon footprints by avoiding burning of residues, e.g., in India in rice monoculture (Yadav et al., 2021).

Integrated nutrient management

Integrated nutrient management (INM) implies the application of chemical fertilizers along with organic resource materials like farmyard manure, composting, green manure, biofertilizers, and other organic decomposable materials which are essential for sustainable crop production.

i *Farmyard manure with chemical fertilizers*: Shahid et al. (2017) reported that the inclusion of farmyard manure enhanced crop yield and carbon sequestration in the long term. Higher carbon sequestration was reported in humid climates of lower Indo Gangetic plains compared to semi-arid climate of the upper Indo Gangetic plain (Nayak et al., 2012). The rate of increase in carbon stock varied between 57 and 89 kg ha^{-1} yr^{-1} by applying chemical fertilizers alone, whereas the rate of increase in carbon stock raised to 61–138 kg ha^{-1} yr^{-1} with the application of chemical fertilizers (NPK) plus farmyard manure. This implies that combined application of chemical fertilizers along with organic materials like farmyard manure increases SOC stocks and sequestration in the soils, which will contribute to reduction in GHG emissions.

ii *Green manuring*: Green manure has been identified as an important substitute to chemical fertilizers in INM practices – particularly applying green manures of leguminous agroforestry trees (e.g., *Gliricidia sepum*: refer Chapter 8 of this book) that provide multiple benefits such as fixing nitrogen and capturing nitrate leaching. In eastern India, green manuring with *Sesbania aculeata* (in direct seeded flood-prone lowland rice) accumulated 80–86 kg N ha^{-1} in pure stand and 58–79 kg N ha^{-1} when intercropped with DSR at 50 days of growth. Dhaincha (*Sesbania aculeata*) manuring was comparable with 40 kg N ha^{-1} as urea in increasing yield of direct seeded and transplanted crops (Sharma and Ghosh, 2000). In general, green manuring with *Sesbania spp.* enhanced labile carbon fractions in soils, thereby enhancing microbial biomass.

iii   *Vermicomposting including biofertilizers:* Vermicomposting is a biological process using a variety of worms including earthworms to transform organic waste into natural nutrient-rich compost, which breaks down organic material through the interaction between worms and microorganisms (Rahman et al., 2019). In India, the application of vermicompost at 20 t ha^{-1} to the soil significantly improved soil porosity and aggregate stability (Ferreras et al., 2006). In rice-legume cropping systems, an integrated application of 50% vermicompost and 50% chemical fertilizer/biofertilizers resulted in 12–20% higher grain yields compared with 100% chemical fertilizer alone (Jeyabal et al., 2001). On-farm demonstration/training on vermicompost preparation and its application was conducted for farmers in Ganjam project areas of Odisha through the Resilience project training programme. Farmers realized the benefits of applying vermicompost into the soils in terms of increase in yields, benefit-cost ratio, and income compared to conventional composting.

Biofertilizers containing blue-green algae with microbial strains such as *Pseudomonas* can potentially substitute part of the nutrient requirement of crops by fixing atmospheric N and provide plant available N. An application of 50% RDN from urea, 25% RDN from blue-green algae, and 25% N from farmyard manure reduced N$_2$O emission by 25% compared to 100% RDN applied Urea alone. Biofertilizer formulation containing *Pseudomonas*, *Penicillium*, and *Actinomycetes species* could enhance total SOC content by 6% after three years of experiment (Debska et al., 2016).

Policy implications

Several schemes have been initiated by the Government of India recently to promote location-specific improved agronomic practices using soil health cards and precision-based tools/techniques.

i   Soil health card scheme
    A major policy measure to improve soil health in India is the Soil Health Card (SHC) scheme which was initiated in 2015 by the Government of India under the Ministry of Agriculture and Farmers' Welfare. The scheme is to ensure soil test-based fertilizer application and improve soil fertility throughout the country. The purpose of the SHC scheme is to provide information to farmers on the soil nutrient status and customize fertilizer management strategy at the farm level and thereby maintain the health of soils. Grids of 10 ha for rainfed and

2.5 ha for irrigated area are used as sampling unit in the SHC scheme. Since 2019–2020, more than 200 million SHCs were distributed to farmers in different states of India. The implementation of the SHC scheme has shown a reduction in urea and DAP fertilizer application by 20–30% in paddy and cotton growing areas. Moreover, it has reduced cost of cultivation by INR 2,500–10,000 ha^{-1} (~US$288–1,150 ha^{-1}). There was also a significant increase in crop yield with adoption of SHC-based fertilizer recommendation and increased farmers' income on average by 30–40% (Reddy, 2018). There is variation in the implementation of the scheme at the state and district levels, with some states in India taking proactive measures, while others still lagging behind.

Site-specific, need-based nutrient management practices and precision tools support the SHC scheme of Indian government by providing evidence-based information/data. This will contribute to improved soil health and conserve the soil ecosystem sustainably. However, uniform grid size adoption for soil testing in the SHC scheme may not represent the field-to-field soil variability. Therefore, grid mapping of soil heterogeneity index using GIS-based soil fertility maps at block/village level is necessary. The index will help to improve precision application of SHC-based fertilizer recommendations.

ii   Support to precision-based tools/techniques

One of the eight national climate missions of India developed under the National Action Plan on Climate Change (NAPCC) was the National Mission for Sustainable Agriculture (NMSA). The NMSA was formulated to enhance agricultural productivity by focusing on several programme components such as soil health management. Under this component, creation of database on soil resources through land use survey, soil profile study, and soil mapping using precision-based tools/techniques (refer to early sections) are one of the key implementation strategies of the NMSA.

The NMSA mission in coordination with other government schemes and missions is promoting specific interventions of precision-based tools/techniques. For example, the LCC technology is supported by NMSA in coordination with other missions and schemes. The precision-based tools/techniques for soil management practices have the potential to increase carbon sequestration and reduce GHG emission. This will contribute to Nationally Determined Contributions (NDCs) and net-zero emission target of India if the initiatives are systematically followed, documented, and reported. It will be necessary to incentivize farmers for generating carbon credits through soil health improvement measures they undertake.

Conclusion

This chapter has presented and discussed some key research results from the ongoing Resilience project to show the positive contributions of precision-based tools/techniques in nutrient management in rice cultivation in Odisha. In addition, soil health improving practices that have the potential to reduce environmental footprints and their policy implications for large-scale adoption were discussed.

i   Precision-based tools/techniques for nutrient management
The precision nutrient and soil management techniques discussed in the chapter have demonstrated advantages in terms of yield advancement, resource saving, enhancement in carbon stock, improvement in soil health, and reducing environmental footprint. Some of the precision nutrient management technologies involve high cost, sophisticated instruments, advanced data collection and analysis. However, data generated using these technologies will be useful for validation and ground toothing while devising large-scale remote sensing-based site-specific recommendation. The advent of remote sensing system such as unmanned air vehicles equipped with multispectral, hyperspectral, and thermal sensors can deliver real-time data at the spatial scale required for precision nutrient management.

The precision-based tools/techniques have considerable adaptation and mitigation potential that could be exploited by providing policy support, adequate funding, infrastructure and extension, and farmer training support as needed. Government policy support and open collaboration with stakeholders at all levels will facilitate adoption of precision soil and nutrient management practices in agriculture.

ii   Soil health improving practices
Agroecological approaches, such as conservation agriculture, with good agronomy and soil management practices such as zero tillage, crop diversifications, and residue mulching have shown an increase in carbon storage and sequestrations, and reduction in GHG emissions and GWP. This is in relation to the emission of methane from rice paddies and nitrogen losses from inefficient use of nitrogen fertilizers. India's commitment for the NMSA – achieving its NDCs, and attaining the goal of zero emission by 2070 – can be supplemented to great extent by promoting cost-effective precision-based tools/techniques with suitable soil health improving practices. In this connection, there is a need to investigate low-cost precision-based tools/techniques for the effective implementation of soil health management practices that are user friendly.

Acknowledgements

The chapter has benefited from the results of the Resilience project field trials in Odisha. The authors would like to thank field enumerators and lead farmers involved in the field demos. At the same time, they acknowledge the support of the Norwegian Ministry of Foreign Affairs/The Norwegian Embassy, New Delhi for funding and support to the Resilience project in India (2018–2023).

References

Alam, M.M., Ladha, J.K., Foyjunnessa Rahman, Z., Khan, S.R., Khan, A.H. and Buresh, R.J. (2006) 'Nutrient management for increased productivity of rice-wheat cropping system in Bangladesh', *Field Crops Res*, vol. 96, 374–386.

Ali, A.M., Thind, H.S., Sharma, S. and Singh, Y. (2015) 'Site-specific nitrogen management in dry direct-seeded rice using chlorophyll meter and leaf color chart', *Pedosphere*, vol. 25(1), 72–81.

Ali, M.M., Al-Ani, A., Eamus, D. and Tan, D.K. (2017) 'Leaf nitrogen determination using non-destructive techniques–A review', *Journal of Plant Nutrition*, vol. 40(7), 928–953.

Al-Kaisi, M.M. and Yin, X. (2005) 'Tillage and crop residue effects on soil carbon and carbon dioxide emission in corn–soybean rotations', *Journal of Environmental Quality*, vol. 34(2), 437–445.

Andrés, P.B., Rozas, H.R.S. and Echeverría, E.H. (2018) 'Nitrogen recovery efficiency from urea treated with NSN co-polymer applied to no-till corn', *Agronomy*, vol. 8 (0154), 1–12.

Bhatia, A., Pathak, H., Jain, N., Singh, P.K. and Tomer, R. (2012) 'Greenhouse gas mitigation in rice–wheat system with leaf color chart-based urea application', *Environmental Monitoring and Assessment*, vol. 184(5), 3095–3107.

Bijay-Singh (2018) 'Are nitrogen fertilizers deleterious to soil health? A review', *Agronomy*, vol. 8, (48), 1–19.

Bijay-Singh (2017) 'Management and use efficiency of fertilizer nitrogen in production of cereals in India: issues and strategies,' In: The Indian Nitrogen Assessment, Sources of Reactive Nitrogen, Environmental and Climate Effects, Management Options, and Policies, eds Y.P. Abrol, T.K. Adhya, V.P. Aneja, N. Raghuram, H. Pathak, U. Kulshrestha, et al., Cambridge: Elsevier, 9–28.

Bijay-Singh, Varinderpal-Singh, Purba, J., Sharma, R.K., Jat, M.L., Yadvinder-Singh, Thind, H.S., Gupta, R.K., Chaudhary, O.P., Chandna, P., Khurana, H.S., Kumar, A., Jagmohan-Singh, Uppal, H.S., Uppal, R.K., Vashistha, M. and Gupta, R. (2015) 'Site-specific fertilizer nitrogen management in irrigated transplanted rice (*Oryza sativa*) using an optical sensor', *Precision Agriculture*, vol. 1, 1–21.

Bijay-Singh, Singh, Y. and Bains, J.S. (2003) 'Real-time nitrogen management using chlorophyll meter and leaf color chart (LCC) in rice and wheat' In: *Nutrient management for sustainable rice-wheat cropping system*, pp. 115–124. NATP, ICAR, New Delhi India and Punjab Agricultural University, Ludhiana.

Buresh, R.J., Castillo, R., van den Berg, M., and Gabinete, G. (2014) 'Nutrient management decision tool for small-scale rice and maize farmers'. *Technical Bulletin 190, Food and Fertilizer Technology Center*, Taipei, Taiwan.

Cassman, K.G., Dobermann, A. and Walters, D.T. (2002) 'Agroecosystems, nitrogen-use efficiency, and nitrogen management', *Ambio*, vol. 31, 132–140.

Dash, P.K., Bhattacharyya, P., Shahid, M., Roy, K.S., Swain, C.K., Tripathi, R. and Nayak, A.K. (2017) 'Low carbon resource conservation techniques for energy savings, carbon gain and lowering GHGs emission in lowland transplanted rice', *Soil Tillage Research*, vol. 174, 45–57.

Dash, P.K. (2019) 'Carbon dynamics and associated diversities under resource conservation technologies in tropical low land rice', PhD Thesis, Utkal University, Odisha.

Debska, B., Długosz, J., Piotrowska-Długosz, A., et al. (2016) 'The impact of a bio-fertilizer on the soil organic matter status and carbon sequestration-results from a field-scale study', *J Soils Sediments*, vol. 16, 2335–2343.

EC (European Commission) (2021) Soil Deal mission implementation plan, section 8B, available at: https://ec.europa.eu/info/publications/implementation-plans-eu-missions_en

FAI (Fertiliser Association of India) (2018) *Fertilizer Statistics* 2017–18, 10th edn. New Delhi.

Ferreras, L., Gomez, E., Toresani, S., Firpo, I. and Ro-tondo, R. (2006) 'Effect of organic amendments on some physical, chemical and biological properties in a horticultural soil', *Bioresource Technology*, vol. 97, 635–640.

Ghosh, M., Swain, D.K., Jha, M.K. and Tewari, V.K. (2013) 'Precision nitrogen management using chlorophyll meter for improving growth, productivity and N use efficiency of rice in subtropical climate', *Journal of Agricultural Sciences*, vol. 5, 253–266.

Huuskonen, J. and Oksanen, T. (2018) 'Soil sampling with drones and augmented reality in precision agriculture', *Computers and Electronics in Agriculture*, vol. 154, 25–35.

IFA (International Fertilizer Association) (2009) 'The global "4R" nutrient stewardship framework: developing fertilizer best management practices for delivering economic, social and environmental benefits,' Paper drafted by the IFA Task Force on Fertilizer Best Management Practices. IFA, Paris, France.

IRRI (International Rice Research Institute) (2019) Annual Progress Report., Increasing productivity of rice–based systems and farmers income in Odisha, India Office, IRRI, 172p

Jahan, N.A., Yeasmin, S., Anwar, M.P., Islam, M.A., Rahman, H. and Islam, A.M. (2018) 'Efficacy and economics of different need-based nitrogen management approaches in winter rice', *American Journal of Plant Sciences*, vol. 9 (13), 2601–2611.

Jat, M.L., Majumdar, K., McDonald, A., Sikka, A.K. and Paroda, R.S. (2015) 'Book of extended summaries, National Dialogue on Efficient Nutrient Management for Improving Soil Health', New Delhi, India, pp. 1–56.

Jeyabal, A. and Kuppuswamy, G. (2001) 'Recycling of organic wastes for the production of vermicompost and its response in rice legume cropping system and soil fertility', *European Journal of Agronomy*, vol. 15, 153–170.

Kumar, P.P. and Sharma, P.K. (2020) 'Soil Salinity and Food Security in India', *Front. Sustain. Food Syst.*, vol. 4 (533781), 1–15.

Kumar, P.P., Abraham, T., Pattanaik, S.S.C., Kumar, R., Kumar, U. and Kumar, A. (2018) 'Effect of customized leaf color chart (CLCC) based real time N management on agronomic attributes and protein content of rice (Oryza sativa L.)', *ORYZA-An International Journal on Rice*, vol. 55(1), 165–173.

Lal, B., Gautam, P., Panda, B.B., Tripathi, R., Shahid, M., Bihari, P., Guru, P.K., Singh, T., Meena, R.L. and Nayak, A.K., (2020) 'Identification of energy and carbon efficient cropping system for ecological sustainability of rice fallow', *Ecological Indicators*, vol. 115 (106431), 1–11.

Lal, B., Priyanka, G., Nayak, A.K., Panda, B.B., Bihari, P., Tripathi, R., Shahid, M., Guru, P.K., Chatterjee, D., Kumar, U. and Meena, B.P. (2019) 'Energy and carbon budgeting of tillage for environmentally clean and resilient soil health of rice-maize cropping system', *Journal of Cleaner Production*, vol. 226, 815–830.

Meena, A.K., Singh, D.K., Pandey, P.C. and Nanda, G. (2018) 'Growth, yield, economics, and nitrogen use efficiency of transplanted rice (*Oryza sativa* L.) as influenced by different nitrogen management practices through neem (*Azadirachta indica*) coated urea', *International Journal of Chemical Studies*, vol. 6(3), 1388–1395.

Mohanty, S., Nayak, A.K., Bhaduri, D., Swain, C.K., Kumar, A., Tripathi, R., Shahid, Md., Behera, K.K. and Pathak, H. (2021) 'Real-time Application of Neem-coated Urea for enhancing N-use efficiency and minimizing Yield Gap between Aerobic Direct-seeded and Puddled Transplanted Rice', *Field Crop Research*, vol. 264(108072), 1–12.

Mohanty, S., Swain, C.K., Tripathi, R., Sethi, S.K., Bhattacharyya, P., Kumar, A., Raja, R., Shahid, M., Panda, B.B., Lal, B., Gautam, P., Munda, S. and Nayak, A.K. (2018) 'Nitrate leaching, nitrous oxide emission and N use efficiency of aerobic rice under different N application strategy', *Archives of Agronomy and Soil Science*, vol. 64, 465–479.

Mohanty, S., Swain, C.K., Sethi, S.K., Dalai, P.C., Bhattachrayya, P., Kumar, A., Tripathi, R., Shahid, M., Panda, B.B., Kumar, U., Lal, B., Gautam, P., Munda, S. and Nayak, A.K. (2017) 'Crop establishment and nitrogen management affect greenhouse gas emission and biological activity in tropical rice production', *Ecological Engineering*, vol. 104, 80–98.

Nayak, A.K., Tripathi, R., Mohanty, S., Shahid, Md., Besra, B., Panda, B.B., Priyadarsani, S., Mohapatra, S.D., Sarangi, D.R., Saha, S., Kar, S., Swain, R., Rajak, M., Moharana,

K.C., Swain, C.K., Nagothu, U.S. and Pathak, H. (2021) 'riceNxpert: A real time nitrogen application tool for enhancing N use efficiency of rice', *NRRI Technology Bulletin 162*, ICAR-National Rice Research Institute, Cuttack, pp. 1–4.

Nayak, A.K., Mohanty, S., Chatterjee, D., Bhaduri, D., Khanam, R., Shahid, M., Tripathi, R., Kumar, A., Munda, S., Kumar, U. and Bhattacharyya, P. (2018) 'Nutrient management for enhancing productivity and nutrient use efficiency in rice', ICAR-NRRI.

Nayak, A.K., Mohanty, S., Chatterjee, D., Guru, P.K., Lal, B., Shahid, M., Tripathi, R., Gautam, P., Kumar, A., Bhattacharyya, P., Panda, B.B., and Kumar, U. (2017) 'Placement of urea briquettes in lowland rice: an environment-friendly technology for enhancing yield and nitrogen use efficiency', *Technical Bulletin No. 12*. ICAR-National Rice Research Institute, Cuttack, pp. 1–26.

Nayak, A.K., Mohanty, S., Raja, R., Shahid, M., Lal, B., Tripathi, R., Bhattacharyya, P., Panda, B.B., Gautam, P., Thilagam, V.K. and Kumar, A. (2013) 'Customized leaf color chart for nitrogen management in rice for different ecologies', CRRI Technology leaflet, Cuttack, India.

Nayak, A.K., Gangwar, B., Shukla, A.K., Mazumdar, S.P., Kumar, A., Raja, R., Kumar, A., Kumar, V., Rai, P.K. and Mohan, U. (2012) 'Long-term effect of different integrated nutrient management on soil organic carbon and its fractions and sustainability of rice–wheat system in Indo Gangetic Plains of India', *Field Crops Research*, vol. 127, 129–139.

Oertel, C., Matschullat, J., Zurba, K., Zimmermann, F. and Erasmi, S. (2016) 'Greenhouse gas emissions from soils—A review', *Chemie der Erde*, vol., 76, 327–352.

Panda, D., Nayak, A.K. and Mohanty, S. (2019) 'Nitrogen management in rice', *Oryza*, vol. 56 (5), 125–135.

Patil, V.B., Desai, B.K. and Jagadish. (2018) 'Effect of nitrogen management through LCC and SPAD meter on growth and yield in direct seeded rice (Oryza sativa L.)', *International Journal of Chemical Studies*, vol. 6(4), 1074–1078.

Peng, S., Buresh, R.J., Huang, J., et al. (2010) 'Improving nitrogen fertilization in rice by site specific N management. A review', *Agronomy for Sustainable Development*, vol. 30, 649–656.

Porpavai, S., Devasenapathy, P., Siddeswaran, K. and Jayaraj, T. (2011) 'Impact of various rice-based cropping systems on soil fertility', *Journal of Cereals and Oilseeds*, vol. 2(3), 43–46.

Rahman, S. (2015) 'Green revolution in India: Environmental degradation and impact on livestock', *Asian Journal of Water, Environment and Pollution*, vol. 12, 75–80.

Rahman, S. and Barmon, B.K. (2019) 'Greening modern rice farming using vermicompost and its impact on productivity and efficiency: an empirical analysis from Bangladesh', *Agriculture*, vol. 9 (239), 1–13.

Reddy, A.A. (2018) 'Impact study of soil health card scheme', National Institute of Agricultural Extension Management (MANAGE), Rajendranagar, Hyderabad—500030, Telangana State, India, 106 pp, available at: https://ssrn.com/abstract=3249953

Sapkota, T.B., Majumdar, K., Khurana, R., Jat, R.K., Stirling, C.M. and Jat, M.L. (2016) 'Precision nutrient management under conservation agriculture-based cereal systems in South Asia'. In: Nagothu, U.S (ed.) *Climate Change and Agricultural Development: Improving Resilience through Climate Smart Agriculture, Agroecology and Conservation*. New York: Routledge, pp: 131–160, ISBN: 978-1-138-92227-3.

Sarker, M.M.R., Shaheb, M.R. and Nazrul, M.I. (2012) 'Urea Super Granule: A Good Source of Nitrogen on Growth Yield and Profitability of Cabbage in Sylhet', *Journal Environmental Science and Natural Resources*, vol. 5(1), 295–299.

Schröder, J.J., Neeteson, J.J., Oenema, O. and Struik, P.C. (2000) 'Does the crop or the soil indicate how to save nitrogen in maize production? Reviewing the state of the art, *Field Crops Research*, vol. 66(2), 151–164.

Sen, A., Srivastava, V.K., Singh, M.K., Singh, R.K. and Kumar, S. (2011) 'Leaf color chart vis-a-vis nitrogen management in different rice genotypes', *American Journal of Plant Sciences*, vol. 2(2), p. 223.

Shahid, M., Nayak, A.K., Puree, C., Tripathi, R., Lal, B., Gautam, P., Bhattacharyya, P., Mohanty, S., Kumar, A., Panda, B.B., Kumar, U. and Shukla, A.K. (2017) 'Carbon and nitrogen fractions and stocks under 41 years of chemical and organic fertilization in a sub-humid tropical rice soil', *Soil and Tillage Research*, vol. 170, 136–146.

Sharma, A.R. and Ghosh, A. (2000) 'Effect of green manuring with Sesbania aculeata and nitrogen fertilization on the performance of direct-seeded flood-prone lowland rice', *Nutrient Cycling in Agroecosystems*, vol. 57, 141–153.

Sharma, S., Panneerselvam, P., Castillo, R., Manohar, S., Rajendren, R., Ravi, V. and Buresh, R.J. (2019) 'Web-based tool for calculating field-specific nutrient management for rice in India, *Nutrient Cycling in Agroecosystems*, vol. 113, 21–33.

Shivay, Y.S., Pooniya, V., Prasad, R., Pal, M. and Bansal, R. (2016) 'Sulphur-coated urea as a source of sulphur and an enhanced efficiency of nitrogen fertilizer for spring wheat', *Cereal Research Communications*, vol. 44 (3), 513–523.

Shukla, A.K., Ladha, J.K., Singh, V.K., Dwivedi, B.S., Balasubramanian, V., Gupta, R.K., Sharma, S.K., Singh, Y., Pathay, H., Pandey, P.S., and Yadav, R.L. (2004) 'Calibrating the leaf color chart for nitrogen management in different genotypes of rice and wheat in a systems perspective, *Agronomy Journal*, vol. 96, 1606–1621.

Singh, V.K., Gautam, P., Nanda, G., Dhaliwal, S.S., Pramanick, B., Meena, S.S., Alsanie, W.F., Gaber, A., Sayed, S. and Hossain, A. (2021) 'Soil test based fertilizer application improves productivity, profitability and nutrient use efficiency of rice (*Oryza sativa* L.) under direct seeded condition', *Agronomy*, vol. 11 (1756), 1–17.

Stolte, J., Tesfai, M., Øygarden, L., Kværnø, S., Keizer, J., Verheijen, F., Panagos, P., Ballabio, C., Hessel, R. (ed.) (2016) 'Soil threats in Europe: status, methods, drivers and effects on ecosystem services', Soil Threats in Europe'; EUR 27607

Susilawati, H.L., Setyanto, P., Kartikawati, R. and Sutriadi, M.T. (2019) 'The opportunity of direct seeding to mitigate greenhouse gas emission from paddy rice field', *IOP Conference Series: Earth and Environmental Science*, vol., 393, 012042.

Syakila, A. and Kroeze, C. (2011) 'The global nitrous oxide budget revisited', *Greenhouse Gas Measurement & Management*, vol. 1, 17–26.

Tripathi, R., Nayak, A.K., Raja, R., Shahid, M., Mohanty, S., Lal, B., Gautam, P., Panda, B.B., Kumar, A., and Sahoo, R.N. (2017) 'Site-specific nitrogen management in rice using remote sensing and geostatistics', *Communications in Soil Science and Plant Analysis*, vol. 48(10), 1154–1166.

Verma, P., Akriti, C. and Ladon, T. (2020) 'Site specific nutrient management: A review', *Research Journal of Pharmacognosy and Phytochemistry*, vol. 9 (5), 233–236.

Yadav, D.B., Yadav, A., Vats, A.K., Gill, G. and Malik, R.K. (2021) 'Direct seeded rice in sequence with zero-tillage wheat in north-western India: addressing system-based sustainability issues', *SN Applied Sciences*, vol. 3, 844.

3 Organizational alternate wetting and drying (AWD) irrigation management in rice by water user groups for reducing methane emission and water saving

Kimihito Nakamura, Le Xuan Quang, and Soken Matsuda

Introduction

Paddy rice agriculture needs to respond to two issues relevant to climate change: one is to reduce greenhouse gas (GHG) emissions, and the other is to save water. Paddy fields are an important source of methane (CH_4), which is one of GHGs responsible for global warming. Though nitrous oxide (N_2O) is also one of GHGs released from the agricultural fields (Nishimura, 2004), the carbon dioxide equivalent of N_2O released from rice paddies is lower than that of methane (Quang et al., 2019). Therefore, methane is the focus in this chapter. Wetland soils are the main natural source with an estimated emission of 100–200 Tg·year^{-1}, whereas the other sources are oceans, forest soils, termites, and wild ruminants. Of the anthropogenic emissions, domesticated ruminants (65–100 Tg·year^{-1}) and rice fields (25–150 Tg·year^{-1}) are responsible for 15–40% of the total emissions. Human activities, including the expansion of paddy rice, played an important role in the observed long-term methane trend over the past two millennia (Sapart et al., 2012). Hence, the need to reduce methane emissions from paddy fields is critical for mitigation (Runkle et al., 2019).

The production and consumption of methane in soils is caused by the metabolic activities of soil microorganisms. Soil organic matter is decomposed by a series of microbial activities, and methane is finally produced by methanogenic bacteria under strongly reducing conditions. When redox potential drops to less than −150 mV (Gupta et al., 2021) or −200 mV (Jean and Pierre, 2001), the methanogenesis process starts. After the production of methane, some of it is consumed by methanotrophs in oxidized zones (rhizosphere, lower part of culms, soil-water interface, and submersion water) (Jean and Pierre, 2001), and some of it is released to the atmosphere. In planted rice fields, there are two pathways of methane from soil to atmosphere: ebullition loss by the release of gas bubbles, and plant transport, into the roots by diffusion and conversion to methane gas in the aerenchyma and cortex of rice plants and concurrent release to the atmosphere

DOI: 10.4324/9781003273172-3

through plant micropores (Davamani et al., 2020; Gupta et al., 2021). At the beginning of the crop cycle, when rice plants are small, the main transfer mechanism is bubble formation and vertical movement in the bulk of the soil. Diffusion through the aerenchyma becomes the dominant process, which is responsible for more than 90% of the methane emitted during the reproductive phase of the rice plant (Cicerone and Shetter, 1981; Jean and Pierre, 2001).

The amount of methane released from rice paddies depends on the redox status of the soil, so, if properly managed, it can be controlled by ponding water management. Draining the continuously flooded rice paddies once or more during the rice-growing season would reduce global emissions by 41 Tg CH_4 year^{-1} (Yan et al., 2009). Therefore, rice paddy water management is very important for climate mitigation.

Paddy rice cultivation uses a large amount of water. There is a concern that climate change will increase the frequency of drought risks (Aryal et al., 2020), resulting in the loss of stable production of rice. Though paddy rice is conventionally grown in lowland systems under continuously flooded conditions, rice can be successfully grown with less water by the adoption of new technologies and various water-saving approaches which have been tested and disseminated (Ishfaq et al., 2020). Water-saving irrigation to maintain rice productivity is required with increase in extreme weather events leading to droughts (Bouman and Tuong, 2001). Alternate wetting and drying (AWD) has been attracting the attention of scientists and farmers as one of the promising water-saving management methods.

In this chapter, the focus is on AWD irrigation. The chapter discusses the different options for the organization of irrigation water management for AWD implementation based on experiences from Japan and Vietnam, and the challenges in AWD adoption followed by potential pathways and conditions for upscaling AWD.

Alternate wetting and drying

The AWD was originally proposed by the Irrigated Rice Research Consortium (IRRC) of the International Rice Research Institute (IRRI) as a water management technique for water-scarce areas (Enriquez et al., 2021). In AWD, irrigation is applied intermittently with a period of non-flooding, whereas in traditional cultivation, the paddy is continuously flooded during the cropping period. Thus, in AWD continuous flooding is maintained for about two weeks after rice transplanting and about two weeks before and after the flowering period. The reason for keeping the fields flooded for two weeks after planting is to suppress weeds and to improve seedling growth. The temporary non-flooding during the cropping season is already practiced in several countries, including China, India, and Japan (Richards and Sander, 2014). In Japan, for example, the drying of rice fields is carried out for about two weeks from about one month after transplanting. This is called the midseason drainage. The purpose of the midseason drainage is to suppress the production of hydrogen sulfide and organic acids due to soil reduction, and to create oxidative conditions in the soil, which inhibits root rot, promotes root elongation of the rice plant, and suppresses the development of non-productive tillers (Shindo et al., 2017). Furthermore, by hardening the surface of the rice fields, harvesting can be carried out smoothly using a harvester even if the field is flooded until the late stage of maturity. Insufficient midseason drying leads to excessive number of stems, which in turn leads to lodging, white underdeveloped grains, additionally weak rooting, and a soft soil surface when the rice reaches maturity. If water is drained early before harvesting, the vigor of the rice will decline in the latter half of the maturity period. Furthermore, the intermittent irrigation with alternating supply of water and oxygen is applied after the midseason drainage, except for the period before and after the flowering period, to keep the vigor of the rice plants intact. However, since most Southeast Asian countries customarily use continuous flooding throughout the cropping season, a non-flooding period is effective in saving water.

The AWD method can contribute to reducing methane emissions since the aerobic soil environment is formed due to the non-flooded period and the methane production is suppressed. As shown by Enriquez et al. (2021), studies conducted in various countries have shown that AWD can reduce GHG emissions and irrigation water use without significant yield loss. Yagi et al. (2020) conducted a meta-analysis, which indicated that water management options, including single and multiple drainage approaches such as AWD, significantly reduced methane emissions by 35% as a mean effect size based on 31 region-specific cases selected for the analysis.

The widely publicized standard technique for intermittent irrigation is to install a field water tube (30 cm long, 15 cm in diameter, either plastic or bamboo pipe with drilled holes) to monitor the water level in the paddy plot and to irrigate water when the water level is 15 cm below the surface until the ponding depth is 3–5 cm. This is called the safe AWD (Rejesus et al., 2011). A depth of 15 cm is the standard to avoid rice yield loss. To reduce yield loss, it is important to level the paddy field surface properly and not to allow dry areas during flooding. The area available for AWD increases in the dry season (Sander et al., 2017). In other words, the effect of AWD varies depending on climatic conditions, especially the amount of rainfall.

Challenges for adopting AWD

AWD is based on the knowledge that rice is tolerant to non-flooded conditions (Kürschner et al., 2010), and does not involve major changes in irrigation facilities or additional costs. However, it requires careful planning and managing irrigation schedules. It is a technique that can be implemented if farmers are aware of the AWD advantages and have a strong motivation. It is not easy to change the mindsets of farmers to switch from the conventional continuous flooded irrigation to AWD because of the risks of poor growth due to soil drying, and overgrowth of weeds and pests. Even if water management is seen as a potential measure to reduce GHG emissions and recommended to farmers, the effects cannot be seen in the short term. Hence, it is difficult for farmers and local agencies to feel the direct and visible benefits. In addition, it is necessary for farmers to monitor the water level in each plot and manage water intake as needed, which can be labor intensive when there are many plots to be managed. Therefore, it is not easy to convince farmers to adopt the AWD system.

The incentives for farmers to adopt AWD also depend on the irrigation system, the water availability, and the associated water pricing scheme. In the case of gravity-driven canal irrigation, there is no incentive for farmers to adopt AWD because the water usage fee is generally fixed per cultivated area or in some cases free of charge, and thus there is no change in the compensation for irrigation water saving, unless there is water shortage. Therefore, in a gravity-irrigation system, the number of farmers who adopt AWD tends to be limited. However, in case where water is used by a pump, farmers have an incentive to adopt AWD because it can save the pump fuel or electric costs for drawing water.

In general, water tends to be scarce in the downstream areas of an irrigation scheme or region compared to the upstream areas. Hence, when AWD is implemented in the upstream area, the possibility for the balanced distribution of water to the downstream area increases, which can contribute to reducing conflicts between the upstream and downstream areas (Rejesus et al., 2014). In this case, it is necessary to foster motivation of farmers especially in the upstream area.

For increasing the adoption of AWD, paddy fields suitable for AWD need to be well considered in terms of irrigation conditions, and simultaneously irrigation facilities need to be developed (Yamaguchi et al., 2019). Additionally, adequate support by local government is essential, for example, training and education of farmers. The economic incentives, organizational strengthening, and improvement of the quality of irrigation infrastructure are important for the widespread adoption of AWD (Enriquez et al., 2021).

Organization of irrigation water management and AWD

While individual farmers' understanding and implementation of AWD is important, relying solely on each farmer to implement AWD, especially when there is no or insufficient direct benefit to the farmer, would not be easy for motivating farmers. Therefore, if the irrigation management organization or water user group can systematically distribute water according to the AWD cycles, the feasibility of AWD irrigation management can potentially be increased. Locally based irrigation management organizations such as water user group (WUG) or water user association (WUA) can play a key role in AWD adoption. The members of the water user groups are basically farmers, but in some cases, experts are hired to take on the role of water managers. Such water user groups can support irrigation projects to control water distribution; operate and maintain irrigation facilities (Teamsuwan and Satoh, 2009); and maintain and manage agricultural dams, head works, pumping equipment, water division works, and main and branch irrigation and drainage canals. However, after the water is distributed from the diversion works, the management of intake and drainage of individual plots is generally implemented by individual farmers.

Water user groups are considered important for the realization of participatory water management (PIM). The policy of PIM has been adopted in several countries to stimulate a more productive and self-reliant irrigated agriculture, and an approach where farmers participate in all stages of irrigation development including operation and maintenance (Hamada and Samad, 2011). For examples, the Government of Egypt attempts to solve challenges related to irrigation water shortage due to increases in water demands by strengthening WUA through the PIM policy (Shindo and Yamamoto, 2017). In Korea, the irrigation management transfer from the local government to the rural community corporation is an emerging social issue (Choi et al., 2016). Similarly, in Japan irrigation projects, all the irrigation facilities, including diversion dams, main and lateral canals, are transferred to land improvement districts (LIDs), which are farmers' autonomous irrigation associations with total responsibility for irrigation system management (Kono et al., 2012). Gany et al. (2019) reviewed institutional reforms in the irrigation sector for sustainable agricultural water management undertaken in 14 countries and regions – including Australia, China, India, Indonesia, Iran, Japan, Malaysia, Mexico, Nepal, South Korea, Sudan, Chinese Taipei, Turkey, and Ukraine – focusing on the legal and organizational framework structure including WUAs for water supply services, PIM and management transfer, and public-private partnership. The study showed that water user groups have a significant impact on the local hydrological cycle and environment in terms of managing the supply of water and improving irrigation and water-use efficiency.

There are several good practices for water saving or conservation where water user groups can themselves manage irrigation and drainage water operations and water quality by directly controlling the water supply. The following are two examples of organized water management in the paddy field district area around Lake Biwa, the largest lake in Japan: block rotational irrigation (BRI) and cyclic irrigation (CI).

Block rotational irrigation

Lake Biwa covers an area of 670.25 km^2 and the lake surface is roughly 85 m above sea level. The water of Lake Biwa was utilized for a variety of purposes, including domestic use, industry, agriculture, and power generation (Shiga Prefectural Government, 2014). Water from the lake was used as a source of irrigation for rice cultivation drawn by electrical pumps. In one of the rice growing areas near the lake covering 1,244 ha, BRI was introduced after a severe drought in 1994 that caused water shortage problems. Irrigation pumps were operated and managed by a WUG (LID). The water pumped from the lake was distributed to all areas simultaneously through 13 division works in the pipeline system. After the 1994 drought, LID decided to introduce the BRI to ensure a stable supply of water. In the BRI, the 13 terminal diversion work areas were divided into three blocks so that the command area of each block was almost equal. The operation schedule was managed such that water was delivered alternatively in two blocks on a given day during the period from mid-July to August, when good water supply was required after the midseason drainage. In other words, a block was irrigated for two days and left without irrigation for one day repeatedly. Figure 3.1 shows an example of the change in the amount of irrigation water supply through BRI for two water diversion work areas (as shown for Unit No. 10 and 11). The extent of irrigation water saving depended on the water diversion work area. The water saving was observed in the No. 11 division area, but not in the No. 10 division area. This may be because the amount of water discharged from the intake valves to paddy plots in the No. 11 division area was stable and intense before introducing BRI, while it was unstable for that of the No. 10 diversion area. In the No. 11 diversion work area where the water supply was stable, the farmers did not manage irrigation to paddy plots every day and often left the intake valves open. As a result, the amount of water used decreased in the BRI because the water was automatically shut off for one day due to the operation of the pump by LID in the No. 11 division area. However, in the No. 10 diversion work area where the water supply was unstable, the effect of the BRI was not observed because the water was not readily available. The water supply to paddy plots was automatically turned on and off due to operation of the block rotational irrigation, and as a result automatic intermittent irrigation occurred. More water saving would have been possible by controlling the water supply consciously by the WUG. Such irrigation experiences from Lake Biwa offered important lessons for effective irrigation water management for other areas in Japan.

While it is important to raise individual farmers' awareness of water saving and management, it is difficult to properly manage water on individual paddy plots in countries such as Japan due to the aging of the farmers, the shift to dual occupation (part-time farming), or the increase in the scale of farming. If irrigation canals and drainage channels are maintained collectively, it may also be possible for WUGs to control irrigation water to the individual paddy plots. Thus,

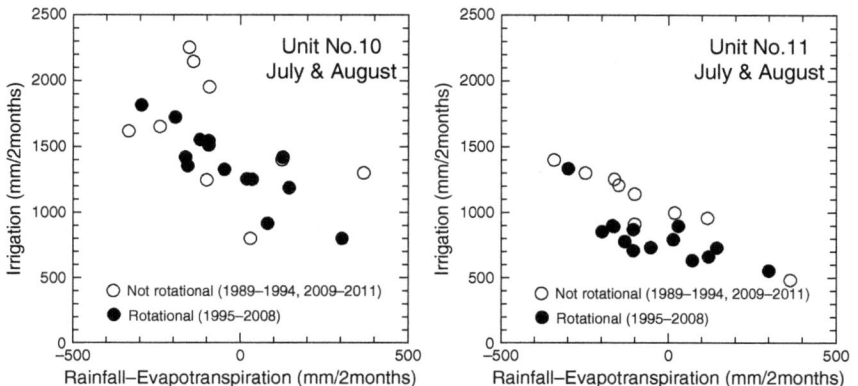

Figure 3.1 Changes in the amount of irrigation by the block rotational irrigation system in the water division work areas around Lake Biwa in Japan. Unit No. 10 is a diversion area with unstable water intake, while No. 11 is an area with stable water intake (authors' own compilation).

WUGs can contribute to improved irrigation water management in each paddy plot through the BRI system.

Cyclic irrigation

Drainage water from rice paddy areas contains nutrients such as nitrogen and phosphorus, especially when the surface water of the rice paddies becomes turbid during the puddling and transplanting seasons in rice paddy cultivation and is discharged until the ponding water level reaches a shallow depth for rice transplanting. Thus, a large amount of nutrient load as well as suspended solids is discharged downstream, affecting the water quality environment and ecosystem of the downstream water body (rivers, lakes, and seas). Therefore, it was important to take measures to reduce the load of nutrients and suspended solids from the paddy fields around Lake Biwa, because the drainage water from the areas adjacent to the lake is immediately discharged back into the lake without undergoing purification or sedimentation. One of the measures suggested was cyclic irrigation, in which water drained from paddy fields was reused as irrigation water. This method was originally developed as a technology to reuse drainage water in areas where water was scarce, but it is now popular in Japan as a means to protect water quality and the environment in general.

In some areas around Lake Biwa, lake water was pumped as the primary source of irrigation water. Therefore, it was possible to install a gate in the drainage channel located near the pumping station and provide a new conduit that diverts the water from the drainage channel back to the irrigation water tank of the pumping station at a relatively low cost. Irrigation water to the command area could thus use water from the lake and the drainage water as well. The concept of the cyclic irrigation system is shown in Figure 3.2. In some districts, in addition to this improvement of water management infrastructures, a new pump only for CI was installed. In one of the areas near the lake, water drainage from a 148 ha area under paddy rice was monitored by Hama et al. (2010). The CI rate, which is the ratio of drainage water to pumped water, was 60–80%. The study showed that the CI system returned 118–199 kg ha^{-1} of suspended solids to the paddy fields, indicating that the runoff loadings were controlled (Hama et al., 2010). Nitrogen concentrations in irrigation water tended to be higher during the CI period than during the lake water irrigation period. Nitrogen input from irrigated water accounted for about 8–16% of the total input of nitrogen (Hama et al., 2011).

The basic management of CI was done by experienced WUGs such as the LID in Lake Biwa, since it involves management of gates of drainage channels, water supply to pumping stations, and management of pumps. This is also important, since farmers will be accepting drainage water from other plots that they do not manage and there is a risk that the entire command area will be affected by inappropriate use and management of pesticides and fertilizers, when drainage water is reused for irrigation. Hence, WUGs can build consensus among all farmers which is necessary in the process and can ensure that all farmers follow certain standard environmental norms. As farmers prefer to take relatively low temperature lake

(a)

(b)

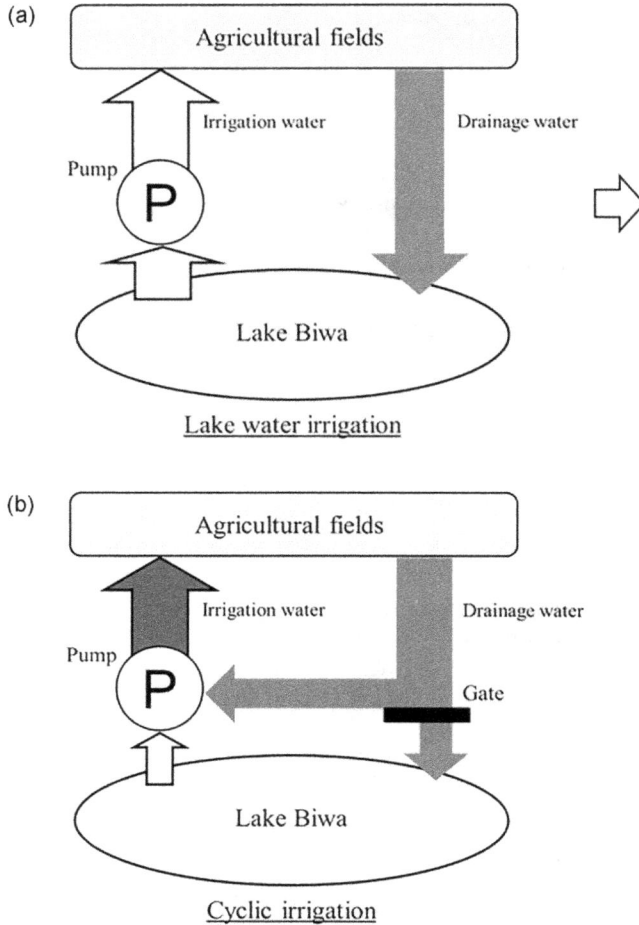

Figure 3.2 From lake water irrigation to cyclic irrigation system (authors' own compilation).

water during the panicle emergence period, CI is generally limited only to puddling and transplanting periods up to the midseason drainage, which could significantly contribute to reduce the amount of runoff and effluent load of nutrients and suspended solids from the paddy field area to outside the district.

Thus, the implementation of water quality protection measures through irrigation and drainage management by WUGs with the consensus of farmers in the paddy field areas can contribute toward better implementation and upscaling of AWD.

Field demonstration of organizational AWD

Using experiences from Japan, studies and demos of AWD management were conducted in the low-lying rice paddies of Kim Dong district located in the Red

River Basin, Hung Yen province of Vietnam. The paddy field area in the command area was about 44 ha. Water was taken from an adjacent canal using pumps
and distributed to the district through an open canal. The pumps were managed
by cooperative agricultural services which is the WUG in the district. The canals
are of dual purpose, used for both irrigation and drainage. The distribution of
water in the district was not controlled, and the pumped water flowed into all
the channels at the same time. The pump operation time was longer and led to
unequal water distribution with overflows in the upper reaches, while there may
not be enough water to reach the fields in the lower reaches. In addition, the area
of a single paddy plot in the district was about 700–1,000 m^2, and it was difficult
to implement AWD management for all plots individually.

Therefore, new diversion works were installed as shown in Figures 3.3 and 3.4 at
two locations along the main irrigation canal. Further, sluice gates were installed
for each of the branching canals. The district was divided into three blocks, and
one block was set as the conventional block, which was always flooded except
the midseason drainage by operating the gates (Block C). The other two blocks
were subjected to AWD conditions: one block was set to irrigate water when the
water level in the paddy plot near the division works was 15 cm below the ground
surface (Block S), and the other block was set to irrigate water when the water
level was 5 cm below the ground surface (Block W). Block W was the safer AWD
condition (weak-dry condition) than Block S (strong-dry condition). Water level
monitoring pipes were installed in the plots near the water division works, and
the water manager operated the pumps and gates by judging from the water level
in the plots. In this area, two seasons' cropping is normally followed for rice: a
winter-spring (WS) crop from February to June and a summer-autumn (SA) crop
from June to September, and various upland crops are grown from October to

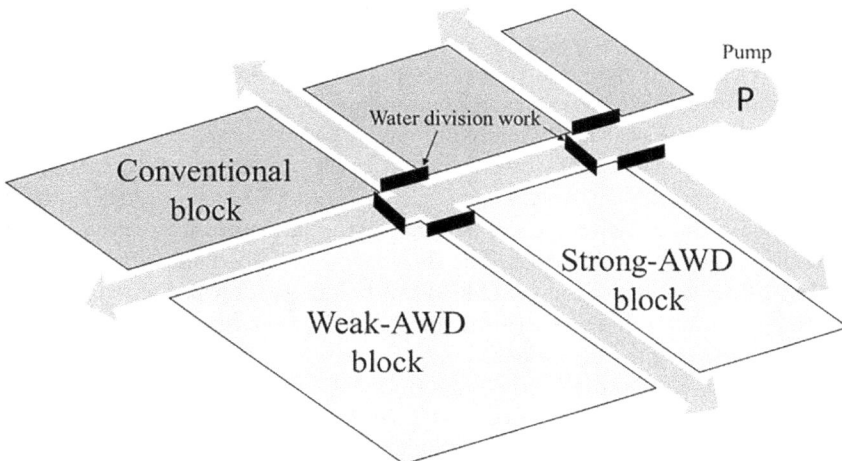

Figure 3.3 Schematic diagram of the investigated paddy area divided into three blocks
with two water division works (authors' own compilation).

Figure 3.4 Water division work with sluice gates for block rotational irrigation. Source: Photo by S. Yonemura.

February. The survey was conducted from the winter-spring crop in 2015 to the summer-autumn crop in 2017.

From the operation records of the pumps and sluice gates, and the temporal changes in the ponding water depth in the observation plots set up in each block, it was clear that the operation was not always ideal depending on the amount and timing of rainfall. Since this was the first experience for water managers, it was difficult for them to operate the system properly, especially in the first year. Figure 3.5 shows the temporal changes in the ponding water depth of the observation plots in 2016. It was observed that water management using organizational AWD method was followed in mid-April during the WS crop season and in July-August during the SA crop season. Figure 3.6 shows the temporal changes in the methane emission fluxes of the observation plots. Comparing with Figure 3.5, the methane emissions were decreased due to the influence of the ponding water depth below the soil surface. The daily average of the cumulative methane emissions in the observed plots divided by the observation period as shown in Figure 3.7 indicated that the organizational AWD suppressed the methane emissions by shortening the period of soil reduction through the ponding water depth control. Figure 3.8 shows that the organizational AWD did not impact yields of paddy rice. Thus, it demonstrated that methane emission could be suppressed without reducing rice yield, if organizational AWD management could be carried out by WUG using BRI to control ponding water depth.

The first challenge in implementing organizational AWD under field condition was that it was difficult for water managers to achieve ideal ponding water depth

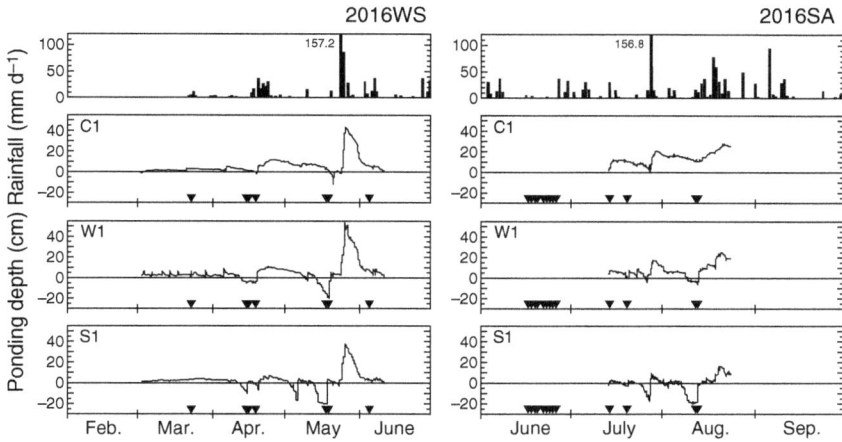

Figure 3.5 Temporal changes in ponding water depths of the rice paddy observation plots in each cropping season in 2016. WS is the winter-spring season and SA is the summer-autumn season. CI, WI, and SI are a conventional plot, weak-dry plot, and strong-dry plot, respectively. Black triangles represent the operation of pumps. Source: adapted from Quang et al. (2019).

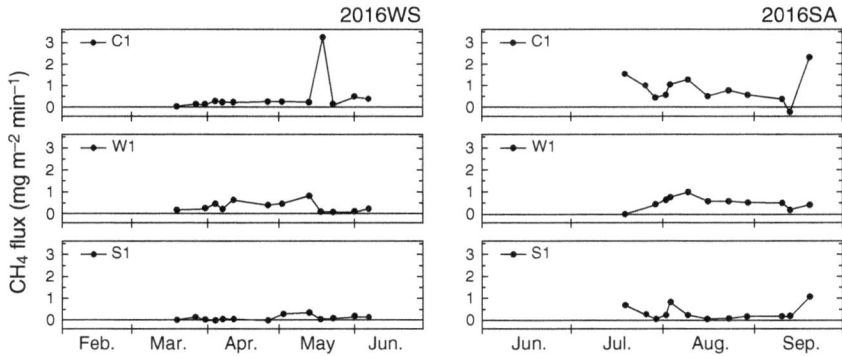

Figure 3.6 Temporal changes in methane emission fluxes in the rice paddy observation plots in each cropping season in 2016. WS is the winter-spring season and SA is the summer-autumn season. CI, WI, and SI are a conventional plot, weak-dry plot, and strong-dry plot, respectively.
Source: adapted from Quang et al. (2019).

management because the depth of water in the plots varied depending on rainfall conditions. During the actual rainfall periods, the efforts to keep the paddy plots under non-flooded condition were reduced. The second problem was that water managers deliberately avoided running irrigation pumps as much as possible to reduce operation costs. In addition, if farmers in the district do not fully understand the implementation of organizational AWD, each farmer may use a small

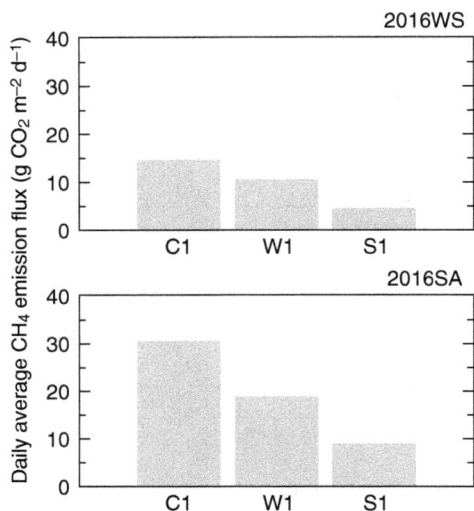

Figure 3.7 Daily average methane emission fluxes converted to the equivalent CO_2 in the observation plots in each cropping season in 2016. WS is the winter-spring season and SA is the summer-autumn season. CI, WI, and SI are a conventional plot, weak-dry plot, and strong-dry plot, respectively. Source: adapted from Quang et al. (2019).

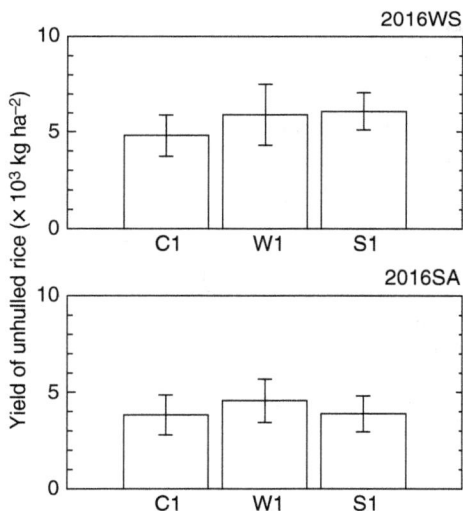

Figure 3.8 Yields of unhulled rice in the observation plots in each cropping season in 2016. Error bars represent the standard deviation. WS is the winter-spring season and SA is the summer-autumn season. C1, W1, and S1 are a conventional plot, weak-dry plot, and strong-dry plot, respectively. Source: adapted from Quang et al. (2019).

manual pump to irrigate water from the canal to each plot when the ponding water depth in his plot is low. The organizational AWD does not work if there is lack of cooperation between water managers and farmers. From the farmers' questionnaire survey, some farmers indicated that they used manual pumps less frequently, suggesting that when the BRI worked, the water level in the canal rose to the downstream in each block compared to the simultaneous water distribution to all plots in the district without the block rotation. Thus, it showed that the BRI with proper operation of pumps and gates at the water division works enabled farmers to realize that they no longer need to operate small manual pumps at the individual plot level. Due to such complexity of the irrigation management, it is likely that there was no clear effect of organizational AWD management on ponding water depth varies and no difference in methane emissions in the cropping seasons, except in 2016. The results of 2016 showed the potential of organizational AWD.

Matsuda et al. (2021) observed the ponding water depth, soil redox potential, and methane emission in 2017 in this district, and found that the methane release could have been suppressed by a continuous non-flooding period of three to eight days after the switch from the flooded state to the non-flooded state, and that the methane was re-emitted after a continuous flooding period of 14–22 days after the switch from the non-flooded state to the flooded state. Therefore, the amount of methane emission can significantly be suppressed by repeating the AWD cycles of non-flooding period of three to eight days and the flooding period of 13–21 days. Such schedules of ponding water depth management for reduction of methane emissions would be useful for water managers. The organizational AWD can be upscaled if such data are measured and made available for different paddy growing regions.

Potential pathways and conditions for upscaling AWD

Multifunctionality and direct payments for environmental services

Demonstration of the organizational AWD in the Red River Basin of Vietnam (Quang et al., 2019) showed that if minor changes were made in irrigation facilities by introducing water diversion facilities, and if pumps and gates of the division works can be properly managed by WUGs, ponding water depth can be managed as needed to suit AWD cycles through the BRI without individual paddy plot water management by individual farmers. The results in such situation showed that the AWD management could significantly contribute to the reduction of methane emissions without major operation costs. In addition, this method enables sufficient water supply for each block, which is expected to avoid the water shortage that has occurred in the downstream areas in the district. To upscale this type of water management to larger areas, it is desirable that water management organizations be financially incentivized or subsidized to implement such environmentally friendly water management. It is also necessary to support the construction costs of necessary minor changes needed in the irrigation facilities, which can contribute to the reduction of GHG emissions for reducing global warming.

Matsuno et al. (2006) reviewed various multifunctionalities of agriculture and the positive impacts of management measures such as flood control, groundwater recharge, soil erosion prevention, landslide prevention, water quality protection, organic waste decomposition, climate mitigation (heat island prevention), biodiversity conservation, landscape formation, and regional development. Some of the major negative impacts include GHG emissions and water pollution of surface water and groundwater due to runoff and infiltration of fertilizer and pesticide components. A study by OECD (2003) indicated that the payment for environmental services (PES) can be justified if the following three conditions can be demonstrated: (1) the agricultural production and multifunctionality are strongly linked; (2) the loss of agricultural production can significantly reduce multifunctionality; and (3) the government support for enhancing multifunctionality and reducing negative impacts is essential. Furthermore, cross compliance is a policy mechanism to encourage farmers to meet certain conditions (e.g., environmental requirements) in return for governmental support payments that had developed in the United States and also gained political attention in the European Union and was developed as a policy instrument (Meyer et al., 2014). This also implies that it is justifiable to provide incentives in terms of financial support to farmers and water management organizations that practice AWD properly to maintain and enhance multifunctionality and in turn can reduce GHG emissions and water pollution.

In Japan, based on the idea that agriculture and rural areas have the multifunctionalities, such as land conservation, water resource recharge, natural environment conservation, landscape formation, and that the benefits of these functions are widely enjoyed by the society at large, the government has introduced three direct payment systems. This is to support local conservation activities and continuation of farming to fulfill the multifunctionalities of agriculture and rural areas, and ensure that the multifunctionalities continue to be properly fulfilled, and to support structural reforms of continuing farmers. The first is the direct payment subsidy for multifunctional activities, which financially supports the collaborative and organizational activities to keep and enhance multifunctionalities and to improve the quality of regional resources. These activities are specifically the basic conservation activities such as mowing the slopes of farmland, the removal of mud in water channels, maintaining the surface of farm roads, extending the service life of agricultural facilities such as water channels, irrigation reservoirs, and others, and landscape formation. The list also includes some specific practices linked to rice cultivation, including the implementation of paddy field dams to enhance flood mitigation (Yoshikawa, 2014), the installation of paddy field fishways for fish-ecosystem conservation (Ohtsuka, 2014), cyclic irrigation system (Hama et al., 2010; Hama et al., 2011), and groundwater recharge from paddy fields (Iwasaki et al., 2014). The PES are provided to organizations composed of farmers and others. The second is the direct payment subsidy for farmers in the hilly and mountainous areas, which supports positive efforts to continue agricultural production activities in the hilly and mountainous areas where the population is aging and declining rapidly. The third is the direct payment for environmentally friendly agriculture, which supports agricultural production activities that are highly effective in reducing the environmental impact of agricultural production, preventing global warming, and conserving biodiversity, and aims to contribute to increasing soil carbon stocks. The requirements for this support include organic farming, cover cropping, composting, no-till sowing of rice, and the prolonged mid-drainage drying in rice cultivations (more than 14 days), which is expected to be effective in reducing methane emissions (Itoh et al., 2011; Kunimitsu and Nishimori, 2020), in addition to reducing the use of chemical fertilizers and synthetic pesticides by at least 50% from the conventional level. Direct payments are made to the organizations by farmers who engage in such activities, as well as to single farmers and corporations with more than a certain cultivated area.

Santos and Shimada (2019) estimated the economic effects of PES for environmentally friendly agriculture on income of rice farmers in Shiga Prefecture, Japan, in which farmers agreed on the following three points: (1) reducing the use of chemical pesticides and fertilizers to less than 50% of the conventional practice, (2) proper use and management of compost and agricultural wastewater, and (3) the agreement period is to be implemented for five years. The study in fact showed that the substitution of environmentally friendly agriculture in place of conventional rice farming had resulted in increased farm income due to the PES received. Kitano (2019) also showed that the direct payment program had a positive influence on the spread of environmental conservation agriculture in Shiga Prefecture.

In Korea, the multifunctionalities of paddy farming have been recognized and the government has introduced a direct payment system for environmentally friendly agriculture to conserve water source and national park areas since 1999, including the minimization of using pesticides and fertilizer and the improvement of water quality for reservoir, stream, and groundwater (Kim et al., 2006). Similarly, Xuehai et al. (2018) concluded that one of the most important approaches to agricultural green development in China is to shift the existing subsidy policy from one which aims to ensure the yield by purchasing at a protective price, to a green subsidy which focuses on agro-ecological compensation. The Vietnamese government passed a national policy on payment for ecosystem services in the hope of strengthening forest conservation, improving local livelihoods, and generating revenue outside of the state budget for nature conservation (To et al., 2012). In Asian countries, the functioning of the PES for agriculture so far seems to be limited to Japan and South Korea.

It is necessary to develop similar support mechanisms in other countries to support farmers that will be responsible for carrying out climate change mitigation and adaptation efforts, including organizational AWD in several countries. Since the reduction of GHG emissions has a global benefit, it is necessary to develop a framework at the global scale beyond the local or national level in the new initiatives to reduce methane from agriculture.

Water management with trade-offs

Agricultural water management affects the soil moisture status, thus altering the soil redox status and chemical and biological reactions of the soil, and the unintended substances are sometimes produced. It has been shown that AWD may increase the emission of nitrous oxide (N_2O), a GHG, which has a greater warming effect than methane (Sibayan et al., 2018). Balaine et al. (2019) mentioned that given the high early season methane fluxes, drying earlier may result in greater reductions of methane in wet seeded rice systems but this requires further study as there may be negative effects such as increased N_2O emissions. Additionally, an increase in the non-flooding period may lead to an oxidative environment and may increase the production of nitrate-nitrogen (NO_3–N), which may leach into groundwater during re-flooding and contribute to an increase in NO_3–N in groundwater. AWD changes the anaerobic environment caused by continuous flooding and accelerates the nitrification process, which improved the consumption of ammonium-nitrogen (NH_4–N) and expedite NO_3–N loading to the groundwater (Wang et al., 2018). The bypass or preferential flow and strengthened nitrification-denitrification nitrogen transformation processes because of AWD potentially increase the NO_3–N loading to the groundwater (Tan et al., 2013). However, Tan et al. (2015) estimated using HYDRUS-1D that the increased NO_3–N, which was formed from nitrification of NH_4–N in drying and aerobic phase, can be easily denitrified to N_2 or N_2O in the wetting and anaerobic phase. In another study, Amin et al. (2021) showed nitrogen leaching depends on the drying spell in AWD irrigation. It will be necessary to study how AWD can be used to reduce the environmental impact of NO_3–N and N_2O.

In some soils, the presence of toxic elements such as arsenic (As) and cadmium (Cd) can be absorbed by crops and become harmful to human health. The mobilities of As and Cd in soils depend on soil redox potential. Paddy rice in flooded soil is prone to As uptake in which reducing conditions increase As mobility through reductive dissolution of As-bearing Fe and Mn oxides and drying cycles create more oxidizing conditions that promote the precipitation of Fe and Mn oxides and subsequent retention of As, which limits As mobility and availability for plant uptake (Evans et al., 2021). However, soil drying may mobilize Cd as sulfide in CdS minerals is oxidized to sulfate and AWD can increase Cd bioavailability (Li et al., 2019). When As and Cd exist simultaneously in paddy soils, the water and soil management must be designed to inhibit crop absorption of both substances. Seyfferth et al. (2019) suggested that a less severe water-saving approach such as AWD in combination with rice residue amendment could limit Cd and As uptake without compromising yield. They mentioned the flooding plus mixed charred/ashed rice husk might limit Cd concentrations in rice and the limited flooding plus Si-rich rice husk might limit As concentrations in rice.

In addition, climate change has already caused severe flood damage, and water storage in rice paddies is expected to be a significant factor in flood prevention in Japan. Efforts to control the amount of drainage from paddy plots as much as possible without damaging the growth of rice are attracting attention (Yoshikawa,

2014). It should be noted that water management is not only about saving water and reducing methane emissions, but also about recycling of nutrients that would otherwise have been lost through drainage water to the surface water, leaching water to the groundwater, or emissions to the atmosphere and other multifunctionalities of paddy rice farming. Therefore, it is necessary to develop more evidence-based scientific knowledge on how to optimize AWD water management with such trade-offs in mind.

Development of necessary information for water management

For an organization to conduct efficient water management, it will be necessary to spatially understand the meteorological conditions, the cropping conditions in the area to be managed, and the characteristics of the field soil, especially the permeability of the soils, which is an important factor that defines the amount of water required. The use of remote sensing (Nagano et al., 2015) and unmanned aerial vehicles can be effective for understanding the status (Krienke et al., 2017; Campo et al., 2020). In addition to the estimations of the spatial distribution of the required water quantity by the remote information systems, it is necessary to comprehensively determine the water use status in the management area and the status of agricultural water utilization facilities. Besides, detailed information on the source water quantity, water level and flow rate in irrigation canals, water level in regulating reservoirs, flow rate in diversion works, groundwater level, and so on is to be regularly recorded in real time.

A variety of hydrological models, which are sometimes coupled with crop model, have been developed for water resources management, agricultural water management, and crop management. Many physically based distributed hydrological models such as the Soil and Water Assessment Tool (SWAT) (Arnold et al., 1993), the Système Hydrologique Européen (SHE) and MIKE SHE (Abbott et al., 1886a, 1886b), Soil and Water Integrated Model (SWIM) (Krysanova et al., 1998), and others have been developed and widely applied. The WEP-L distributed hydrological model, derived from the water and energy transfer processes (WEP), coupled simulations of natural hydrological processes and water use processes by human activities (Jia et al., 2006). Khadim et al. (2021) developed a numerical framework, based on a groundwater model using MODFLOW-NWT (Niswonger et al., 2005), coupled with the outputs of the DSSAT crop model (Jones et al., 2003) for agricultural water management. Furthermore, to apply the hydrological models to Asian regions dominated by paddy fields, for example, Xie and Cui (2011) customized the SWAT model by incorporating new processes for irrigation and drainage.

The development of an integrated hydrological simulation model including irrigation and drainage management in paddy and upland fields, and the operations of agricultural water facilities (dams, head works, reservoir, water division works, groundwater pumps, and so on) in addition to the natural surface-subsurface hydrological processes are necessary. Outputs from such models can provide accurate information that is necessary for water management organizations to operate AWD.

ICT *in water management*

With the advancement of information and communication technology (ICT), it is relatively easy to implement not only remote acquisition of data such as water flow and water level at agricultural water utilization facilities, but also remote operation and control such as opening and closing of gates and valves and starting and stopping of operation. In Korea, smart agricultural water resources management systems have also been introduced in irrigation districts as the prototype projects managing and monitoring the irrigation system from the water resources to the irrigated fields and ICT can be an alternative solution to overcome vulnerability of agricultural water resources structures impacted by droughts and floods (Choi, 2015). A study by Masseroni et al. (2018) evaluated the hydraulic, control, and economical performances of the automatic and remote-controlled system applied for traditional rice irrigation in Italy. The study demonstrated that the automatic system allowed us to drastically reduce the time spent by workers for water level control and flow regulation and the price of the automatic irrigation system appeared to be in good agreement with respect to the willingness of farmers for innovation. Although there are many hurdles to overcome in terms of installation and running costs, it will be useful for organizational AWD in the long term as a labor-saving method if the operation of pumps and gates of diversion works can be automated while monitoring the water level in the canal and the paddy ponding water depth in the plots in the district. Furthermore, an automatic water supply system that starts water intake at a certain time, stops water intake when the water level in the paddy plot reaches the upper limit, and resumes water intake when the water level reaches the lower limit has been developed in paddy fields in Japan. It is expected to facilitate more efficient irrigation management (Nishida et al., 2022). The ICT water management system is technically feasible for AWD management.

Conclusions

To reduce the contribution to GHG emissions from rice paddies, it is important to focus on paddy water management with particular attention to methane reduction. In addition, it is important to have the potential ability for water-saving paddy water management in response to climate change. To spread AWD in a wide area and continuously, the organizational AWD irrigation system by WUGs based on BRI with the operation of water diversion works along the irrigation channels.

Since the reduction of methane emission by AWD contributes to the prevention of global warming, it might be useful to promote AWD by establishing a PES mechanism to benefit farmers at the national level for water management organizations and farmer groups that strictly implement AWD with cross compliance.

Providing an appropriate paddy ponding management schedule considering AWD to water management organizations and farmers is necessary. It should be a specific water and soil management method according to the soil and climatic characteristics of the region, not only considering methane emission and water conservation, but also reducing the environmental load of nitrate nitrogen and nitrous oxide and crop uptake of Cd, which are caused by the trade-off between the oxidative change of soil due to the non-flooding period increased by AWD.

In addition, it is necessary to establish sensing and analysis technologies to obtain information on water consumption, especially from spatial and temporal meteorological and agricultural crop conditions; to construct and operate a hydrological model that can indicate water management for agricultural water use facilities and paddy fields based on the information; and to build a platform to ensure that water management organizations and farmers can easily implement the proposed appropriate water management using ICT.

Acknowledgments

AWD management demos were conducted under a joint initiative of the Institute for Water and Environment under the Vietnam Academy for Water Resources in Vietnam and the Industrial-Academic Cooperative Research for Irrigation Water Management in Japan comprising Kitai Sekkei Co., Ltd. and Kyoto University. The project was supported by the Ministry of Science and Technology in Vietnam (Grant Number: NĐT.06.JPN/15), and the Japan Society for the Promotion of Science (JSPS) KAKENHI (16H05799). We are grateful for the cooperation of the farmers and the agricultural cooperatives involved, the Hinogawa River Basin LID, the Kamogawa River Basin LID, the Konohama LID, the Shiga Prefecture Office, the Kinki Regional Agricultural Administration Office of the Ministry of Agriculture, Forestry and Fisheries, Japan, JSPS KAKENHI (18H02297 and 20H03101), and "Integrated Research Program for Advancing Climate Models (TOUGOU program)" from the Ministry of Education, Culture, Sports, Science and Technology, Japan.

References

Abbott, M.B., Bathurst, J.C., Cunge, J.A., O'Connell, P.E. and Rasmussen, J. (1986a) 'An introduction to the European hydrological system – Systeme Hydrologique Europeen, "SHE", 1: History and philosophy of a physically-based, distributed modelling system', *Journal of Hydrology*, vol. 87, pp. 45–59.

Abbott, M.B., Bathurst, J.C., Cunge, J.A., O'Connell, P.E. and Rasmussen, J. (1986b) 'An introduction to the European hydrological system – Systeme Hydrologique Europeen, "SHE", 2: Structure of a physically-based, distributed modelling system', *Journal of Hydrology*, vol. 87, pp. 61–77.

Amin, M.G.M., Akter, A., Jahangir, M.M.R. and Ahmed, T. (2021) 'Leaching and runoff potential of nutrient and water losses in rice field as affected by alternate wetting and drying irrigation', *Journal of Environmental Management*, vol. 297, 113402.

Arnold, J.G., Allen, P.M. and Bernhardt, G. (1993) 'A comprehensive surface-groundwater flow model', *Journal of Hydrology*, vol. 142, pp. 47–69.

Aryal, J.P., Sapkota, T.B., Khurana, R., Khatri-Chhetri, A., Rahut, D.B. and Jat, M.L. (2020) 'Climate change and agriculture in South Asia: adaptation options in smallholder production systems', *Environment, Development and Sustainability*, vol. 22, pp. 5045–5075.

Balaine, N., Carrijo, D.R., Adviento-Borbe, M.A. and Linquist, B. (2019) 'Greenhouse gases from irrigated rice systems under varying severity of alternate-wetting and drying irrigation', *Soil Science Society of America Journal*, vol. 83, pp. 1533–1541.

Bouman, B.A.M. and Tuong, T.P. (2001) 'Field water management to save water and increase its productivity in irrigated lowland rice', *Agricultural Water Management*, vol. 49, pp. 11–30.

Campo, L.V., Ledezma, A. and Corrales, J.C. (2020) 'Optimization of coverage mission for lightweight unmanned aerial vehicles applied in crop data acquisition', *Expert Systems with Applications*, vol. 149, 113227.

Choi, J.Y. (2015) 'Irrigation and drainage in Korea and ICT applications', *Irrigation and Drainage*, vol. 65, pp. 157–164.

Choi, S.M., Yoon, K.S. and Kim J.S. (2016) 'Irrigation management transfer between public organizations and the role of participatory irrigation management under public irrigation management in Korea', *Irrigation and Drainage*, vol. 65, pp. 69–75.

Cicerone, R.J. and Shetter, J.D. (1981) 'Sources of atmospheric methane: measurements in rice paddies and a discussion', *Journal of Geophysical Research*, vol. 86 (C8), pp. 7203–7209.

Davamani, V., Parameswari, E. and Arulmani, S. (2020) 'Mitigation of methane gas emissions in flooded paddy soil through the utilization of methanotrophs', *Science of the Total Environment*, vol. 726, 138570.

Enriquez, Y., Yadav, S., Evangelista, G.K., Villanueva, D., Burac, M.A. and Pede, V. (2021) 'Disentangling challenges to scaling alternate wetting and drying technology for rice cultivation: Distilling lessons from 20 years of experience in the Philippines', *Frontiers in Sustainable Food Systems*, vol. 5, p. 675818.

Evans, A.E., Limmer, M.A. and Seyfferth, A.L. (2021) 'Indicator of redox in soils (IRIS) films as a water management tool for rice farmers', *Journal of Environmental Management*, vol. 294(15), p. 112920.

Gany, A.H.A., Sharma, P. and Singh, S. (2019) 'Global review of institutional reforms in the irrigation sector for sustainable agricultural water management, including water users' associations', *Irrigation and Drainage*, vol. 68, pp. 84–97.

Gupta, K., Kumar, R., Baruah, K.K., Hazarika, S., Karmakar, S. and Bordoloi, N. (2021) 'Greenhouse gas emission from rice fields: a review from Indian context', *Environmental Science and Pollution Research*, vol. 28, pp. 30551–30572.

Hama, T., Nakamura, K. and Kawashima, S. (2010) 'Effectiveness of cyclic irrigation in reducing suspended solids load from a paddy-field district', *Agricultural Water Management*, vol. 97, pp. 483–489.

Hama, T., Nakamura, K., Kawashima, S., Kaneki, R. and Mitsuno, T. (2011) 'Effects of cyclic irrigation on water and nitrogen mass balances in a paddy field', *Ecological Engineering*, vol. 37, pp. 1563–1566.

Hamada, H. and Samad, M. (2011) 'Basic principles for sustainable participatory irrigation management', *Japan Agricultural Research Quarterly*, vol. 45(4), pp. 371–376.

Ishfaq, M., Farooq, M., Zulfiqar, U., Hussain, S., Akbar, N., Nawaz, A. and Anjum, S.A. (2020) 'Alternate wetting and drying: A water-saving and ecofriendly rice production system', *Agricultural Water Management*, vol. 241, p. 106363.

Itoh, M., Sudo, S., Mori, S., Saito, H., Yoshida, T., Shiratori, Y., Suga, S., Yoshikawa, N., Suzue, Y., Mizukami, H., Mochida, T. and Yagi, K. (2011) 'Mitigation of methane emissions from paddy fields by prolonging midseason drainage', *Agriculture, Ecosystems & Environment*, vol. 141, pp. 359–372.

Iwasaki, Y., Nakamura, K., Horino, H. and Kawashima, S. (2014) 'Assessment of factors influencing groundwater-level change using groundwater flow simulation, considering vertical infiltration from rice-planted and crop-rotated paddy fields in Japan', *Hydrogeology Journal*, vol. 22(8), pp. 1841–1855.

Jean, L.M. and Pierre, R. (2001) 'Production, oxidation, emission and consumption of methane by soils: A review', *European Journal of Soil Biology*, vol. 37, pp. 25–50.

Jia, Y., Wang, H., Zhou, Z., Qiu, Y., Luo, X., Wang, J., Yan, D. and Qin, D. (2006) 'Development of the WEP-L distributed hydrological model and dynamic assessment of water resources in the Yellow River basin', *Journal of Hydrology*, vol. 331, pp. 606–629.

Jones, J.W., Hoogenboom, G., Porter, C.H., Boote, K.J., Batchelor, W.D., Hunt, L.A., Wilkens, P.W., Singh, U., Gijsman, A.J. and Ritchie, J.T. (2003) 'The DSSAT cropping system model', *European Journal of Agronomy*, vol. 18(3-4), pp. 235–265.

Khadim, F.K., Dokou, Z., Bagtzoglou, A.C., Yang, M., Lijalem, G.A. and Anagnostou, E. (2021) 'A numerical framework to advance agricultural water management under hydrological stress conditions in a data scarce environment', *Agricultural Water Management*, vol. 254, p. 106947.

Kim, T.C., Gim, U.S., Kim, J.S. and Kim, D.S. (2006) 'The multi-functionality of paddy farming in Korea', *Paddy Water and Environment*, vol. 4, pp. 169–179.

Kitano, S. (2019) 'An evaluation of a direct payment policy for community-based environmental conservation agricultural practices: A case of Shiga prefecture in Japan', *Journal of Environmental Information Science*, vol. 2019(1), pp. 43–52.

Kono, S., Ounvichit, T., Ishii, A. and Satoh, M. (2012) 'Participatory system for water management in the Toyogawa Irrigation Project, Japan', *Paddy and Water Environment*, vol. 10, pp. 75–81.

Krienke, B., Ferguson, R.B., Schlemmer, M., Holland, K., Marx, D. and Eskridge, K. (2017) 'Using an unmanned aerial vehicle to evaluate nitrogen variability and height effect with an active an active crop canopy sensor', *Precision Agriculture*, vol. 18, pp. 900–915.

Krysanova, V., Muller-Wohlfeil, D.I. and Becker, A. (1998) 'Development and test of a spatially distributed hydrological/water quality model for mesoscale watersheds', *Ecological Modelling*, vol. 106, pp. 261–289.

Kunimitsu, Y. and Nishimori, M. (2020) 'Policy measures to promote mid-summer drainage in paddy fields for a reduction in methane gas emissions: the application of a dynamic, spatial computable general equilibrium model', *Paddy and Water Environment*, vol. 18, pp. 211–222.

Kürschner, E., Henschel, C., Hildebrandt, T., Jülich, E., Leineweber, M. and Paul, C. (2010) 'Water saving in rice production-dissemination, adoption, and short-term impacts of alternate wetting and drying (AWD) in Bangladesh', SLE Publication Series S241, Humboldt Universitat, Berlin.

Li, C., Carrijo, D.R., Nakayama, Y., Linquist, B.A., Green, P.G. and Parikh, S.J. (2019) 'Impact of alternate wetting and drying irrigation on arsenic uptake and speciation in flooded rice systems', *Agriculture, Ecosystems and Environment*, vol. 272, pp. 188–198.

Masseroni, D., Moller, P., Tyrell, R., Romani, M., Lasagna, A., Sali, G., Facchi, A. and Gandolfi, C. (2018) 'Evaluating performances of the first automatic system for paddy irrigation in Europe', *Agricultural Water Management*, vol. 201, pp. 58–69.

Matsuda, S., Nakamura, K., Hung, T., Quang, L.X., Horino, H., Hai, P.T., Ha, N.D. and T. Hama (2021) 'Paddy ponding water management to reduce methane emission based on observations of methane fluxes and soil redox potential in the Red River Delta, Vietnam', *Irrigation and Drainage*, vol. 71(1), pp. 241–254.

Matsuno, Y., Nakamura, K., Masumoto, T., Matsui, H., Kato, T. and Sato, Y. (2006) 'Prospects for multifunctionality of paddy rice cultivation in Japan and other countries in monsoon Asia', *Paddy Water Environment*, vol. 4, pp. 189–197.

Meyer, C., Matzdorf, B., Müller, K. and Schleyer, C. (2014) 'Cross Compliance as payment for public goods? Understanding EU and US agricultural policies', *Ecological Economics*, vol. 107, pp. 185–194

Nagano, T., Ono, Y., Kotera, A. and Singh, R. (2015) 'Detecting fluctuation of rice cultivated areas in semi-arid regions by combined use of MODIS and Landsat imageries', *Hydrological Research Letters*, vol. 9(4), pp. 107–112.

Nishida, K., Yoshida, S. and Shiozawa, S. (2022) 'Numerical simulation on effect of irrigation conditions on water temperature distribution in a paddy field', *Paddy and Water Environment*, published online, https://doi.org/10.1007/s10333-021-00884-1.

Nishimura, S., Sawamoto, T., Akiyama, H., Sudo, S. and Yagi, K. (2004) 'Methane and nitrous oxide emissions from a paddy field with Japanese conventional water management and fertilizer application', *Global Biogeochemical Cycles*, vol. 18, GB2017.

Niswonger, R.G., Panday, S. and Ibaraki, M. (2011) 'MODFLOW-NWT, A Newton formulation for MODFLOW-2005', U.S. Geological Survey Techniques and Methods, 6-A37, 44 p.

OECD (2003) 'Multifunctionality: The Policy Implications', OECD.

Ohtsuka, T. (2014) 'Nursery grounds for round crucian carp, *Carassius auratus grandoculis*, in rice paddies around lake Biwa', In: Usio, N. and Miyashita, T. (Eds.), *Social-Ecological Restoration in Paddy-Dominated Landscapes*, Ecological Research Monographs, Springer, Tokyo.

Quang, L.X., Nakamura, K., Hung, T., Tinh, N.V., Matsuda, S., Kadota, K., Horino, H., Hai, P.T., Komatsu, H., Hasegawa, K., Fukuda, S., Hirata, J., Oura, N., Kishimoto-Mo, A.W., Yonemura, S. and Onishi, T. (2019) 'Effect of organizational paddy water management by a water user group on methane and nitrous oxide emissions and rice yield in the Red River Delta, Vietnam', *Agricultural Water Management*, vol. 217, pp. 179–192.

Rejesus, R.M., Martin, A.M. and Gypmantasiri, P. (2014) 'Enhancing the impact of natural resource management research: lessons from a meta-impact assessment of the Irrigated Rice Research Consortium', *Global Food Security*, vol. 3, pp. 41–48.

Rejesus, M.R., Palis, F.G., Rodriguez, D.G., Lampayan, R.M. and Bouman, B.A.M. (2011) 'Impact of the alternate wetting and drying (AWD) water-saving irrigation technique: evidence from rice producers in the Philippines', *Food Policy*, vol. 36, pp. 280–288.

Richards, M. and Sander, B.O. (2014) 'Alternate wetting and drying in irrigated rice', Climate-Smart Agriculture Practice Brief, Copenhagen, Denmark: CGIAR Research Program on Climate Change, Agriculture and Food Security (CCAFS).

Runkle, B.R.K., Suvočarev, K., Reba, M.L., Reavis, C.W., Smith, S.F., Chiu, Y.L. and Fong, B. (2019) 'Methane emission reductions from the alternate wetting and drying of rice fields detected using the eddy covariance method', *Environmental Science & Technology*, vol. 53, pp. 671–681.

Sander, B.O., Wassmann, R., Palao, L.K. and Nelson, A. (2017) 'Climate-based suitability assessment for alternate wetting and drying water management in the Philippines: a novel approach for mapping methane mitigation potential in rice production', *Carbon Management*, vol. 8:4, pp. 331–342.

Santos, D.K. and Shimada, K. (2019) 'Analysis of the effects of direct payment subsidies for environmentally-friendly agriculture on income of rice farmers in Shiga, Japan', *European Journal of Sustainable Development*, vol. 8(4), pp. 325–336.

Sapart, C.J., Monteil, G., Prokopiou, M., van de Wal, R.S.W., Kaplan, J.O., Sperlich, P., Krumhardt, K.M., van der Veen, C., Houweling, S., Krol, M.C., Blunier, T., Sowers, T., Martinerie, P., Witrant, E., Dahl-Jensen, D. and Röckmann, T. (2012) 'Natural and anthropogenic variations in methane sources during the past two millennia', *Nature*, vol. 490, pp. 85–88.

Seyfferth, A.L., Amaral, D., Limmer, M.A. and Guilherme, L.R.G. (2019) 'Combined impacts of Si-rich rice residues and flooding extent on grain as and Cd in rice', *Environment International*, vol. 128, pp. 301–309.

Shindo, H., Saito, M., Sasaki, K. Sato, Y. and Katahira, M. (2017) 'Characteristics of seedbed preparation work using a chisel plow and power harrow and rice plant growth on non-puddling direct rice seeding cultivation in flooded paddy field soil', *Japanese Journal of Farm Work Research*, vol. 52(3), pp. 121–131. (in Japanese with English abstract)

Shindo, S. and Yamamoto, K. (2017) 'Strengthening water users' organization targeting for participatory irrigation management in Egypt', *Paddy and Water Environment*, vol. 15, pp. 773–785.

Sibayan, E., Samoy-Pascual, K., Grospe, F., Casil, M., Tokida, T., Padre, A. and Minamikawa, K. (2018) 'Effects of alternate wetting and drying technique on greenhouse gas emissions from irrigated rice paddy in Central Luzon, Philippines', *Soil Science and Plant Nutrition*, vol. 64, pp. 39–46.

Shiga Prefectural Government (2014) 'Lake Biwa Guidebook', Environmental Policy Division of Shiga Prefectural Government, 131 p, Available at: https://www.pref.shiga.lg.jp/ippan/kankyoshizen/biwako/13473.html

Tan, X., Shao, D., Liu, H., Yang, F., Xiao, C. and Yang H. (2013) 'Effects of alternate wetting and drying irrigation on percolation and nitrogen leaching in paddy fields', *Paddy and Water Environment*, vol. 11, pp. 381–395.

Tan, X., Shao, D., Gu, W. and Liu, H. (2015) 'Field analysis of water and nitrogen fate in lowland paddy fields under different water managements using HYDRUS-1D', *Agricultural Water Management*, vol. 150, pp. 67–80.

Teamswan, V. and Satoh, M. (2009) 'Comparative analysis of management of three water users' organizations: successful cases in the Chao Phraya Delta, Thailand', *Paddy and Water Environment*, vol. 7, pp. 227–237.

To, P.X., Dressler, W.H., Mahanty, S., Pham, T.T. and Zingerli, C. (2012) 'The prospects for payment for ecosystem services (PES) in Vietnam: A look at three payment schemes', *Human Ecology*, vol. 40, pp. 237–249.

Wang, M., Yu, S., Shao, G., Gao, S., Wang, J. and Zhang, Y. (2018) 'Impact of alternate drought and flooding stress on water use, and nitrogen and phosphorus losses in a paddy field', *Polish Journal of Environmental Studies*, vol. 27(1), pp. 345–355.

Xie, X. and Cui, Y. (2011) 'Development and test of SWAT for modeling hydrological processes in irrigation districts with paddy rice', *Journal of Hydrology*, vol. 396(1–2), pp. 61–71.

Xuehai, J., Yinghao, X., Bin, X., Tuo, J., Zhiyu, X. and Shangbin, G. (2018) 'Establishing an agro-ecological compensation mechanism to promote agricultural green development in China,' *Journal of Resources and Ecology*, vol. 9(4), pp. 426–433.

Yagi, K., Sriphirom,P., Cha-un, N., Fusuwankaya, K., Chidthaisong, A., Damen, B and Towprayoon, S. (2020) 'Potential and promisingness of technical options for mitigating greenhouse gas emissions from rice cultivation in Southeast Asian countries', *Soil Science and Plant Nutrition*, vol. 66(1), pp. 37–49.

Yamaguchi, T., Tuan, L.M., Minamikawa, K. and Yokoyama, S. (2019) 'Assessment of the relationship between adoption of a knowledge-intensive water-saving technique and irrigation conditions in the Mekong Delta of Vietnam', *Agricultural Water Management*, vol. 212, pp. 162–171.

Yan, X., Akiyama, H., Yagi, K. and Akimoto, H. (2009) 'Global estimations of the inventory and mitigation potential of methane emissions from rice cultivation conducted using the 2006 Intergovernmental Panel on Climate Change Guidelines', *Global Biogeochemical Cycles*, vol. 23, p. GB2002.

Yoshikawa, N. (2014) 'Can paddy fields mitigate flood disaster? Possible use and technical aspects of the paddy field dam', In: Usio, N. and Miyashita, T. (Eds), *Social-Ecological Restoration in Paddy-Dominated Landscapes*, Ecological Research Monographs, Springer, Tokyo.

4 Integrated pest management in rice and the potential to contribute to climate-neutral and resilient farming systems

Shyamaranjan Das Mohapatra, Najam Waris Zaidi, Minati Mohapatra, Udaya Sekhar Nagothu, Radhakrushna Senapati, Munmun Mohapatra, Bhubananda Adhikari, Subhendu Sekhar Pradhan, and Amaresh Kumar Nayak

Introduction

It is now 60 years since the book *The Silent Spring* was published in 1962 by Rachel Carson, based on experiences of communities using persistent pesticides, which caused serious damage to biodiversity, pollinators, wildlife, farm livestock, and human health (NRDC, 2015). The book raised alarm about the negative impacts of chemical pesticides on the environment and the future of the planet. Her message is highly relevant even today, with the current climate crisis we are facing, where agriculture is becoming both a victim and cause for climate crisis. Although several measures were taken by governments to ban the use of toxic pesticides, their use still continues in several parts of the world. The overall impact of intensive agriculture promoted by the Green Revolution in the 1960s and 1970s, and the indiscriminate use of agrochemicals in the following years led to irreparable damage to agroecosystems, the biodiversity they harbor, and an increase in greenhouse gases (GHGs) as discussed in Chapter 1. *The Silent Spring* and other similar publications triggered environmental debates around the globe that gradually increased awareness among scientists, stakeholders, and development agencies about the importance of sustainable and integrated pest management (IPM) with minimum impacts on the environment and human health.

Changes in temperature, precipitation patterns, humidity, and droughts have variable impacts on the different pest and disease species and their behavior, which can become unpredictable with more increased frequency of extreme weather events. One direct impact that is of concern to agriculture is increased incidence and distribution of insect pests, diseases, and weeds, which leads to low crop production and productivity (Oerke, 2006). Climate change is affecting the outbreak of potential pests in a vast range of crops and landscapes (Heeb et al., 2019). Due to the combined effect of climate change and unsustainable agronomic practices, new invasive species occur, while some of the minor pests

DOI: 10.4324/9781003273172-4

and diseases are becoming a major problem (Mohapatra et al., 2008, 2016, 2021b; Pretty and Bharucha, 2015).

A recent study revealed that there was an increase in the duration of the developmental period of yellow stem borer, *Scirpophaga incertulas*, in each stage of the life cycle as the concentration of CO_2 increases (Giri et al., 2022). However, the study showed that the life span of the adult moth was significantly reduced under elevated CO_2 concentrations when compared with ambient CO_2 concentrations (Giri et al., 2022). Similarly, the Fall Army Worm outbreak in Africa and other regions impacted by global warming has drawn the attention of several national governments and international agencies to jointly develop an IPM framework to check the spread of pests (Prasanna et al., 2018). There are several other examples of such effects of rise in average temperature on other pests and diseases. Therefore, future management strategies must consider not only the pest incidence and control *per se* but also the climate and environmental risks. In this chapter, the focus will be on climate-sensitive and sustainable IPM practices in rice, which is the major staple food for more than 60% of the population and of crucial importance to global food security (Yuan, 2014).

Pest and disease scenarios and dynamics of IPM in rice-growing regions of Asia

Pests and diseases cause both direct and indirect losses to crops. The direct crop yield losses roughly account for 20–40% (Teng, 1987; Oerke et al., 1996; Oerke, 2006). In rice, diseases such as rice blast, rice bacterial blight, and false smut, and insect pests such as stem borers and brown planthopper (BPH) are the major causes for crop damage. In addition, we are witnessing new invasive pests and diseases entering new areas through import of seed/planting materials with the increase in global trade of agricultural products, despite the quarantine regulations in place (European Commission, 2016).

As farming practices evolved over time, associated pests have adapted to them. Since the introduction of high yielding varieties during the Green Revolution, combined with prophylactic use of chemical pesticides, the number of insect pests and diseases also increased with a corresponding change in their intensity, diversity, and distribution in rice crop (John and Babu, 2021). Large-scale outbreaks of insect pests have occurred in the past, because of pesticide subsidies and subsequent overuse in some countries (Kenmore, 1996). A study by Rao and Rao (2017) in India showed that 73% of farmers sprayed chemical pesticides as soon as they noticed the first appearance of pests, irrespective of the need. Another study from the Philippines reported that 80% of rice farmers' pesticide sprays were not used properly (Heong et al., 1995). Both insecticides and fertilizers then were subsidized by governments and international Overseas Development Programs, some of which continue even today. The negative impacts of chemical pesticides on human health, food safety, and the ecosystem, including reduced biodiversity and ecosystem services were well documented in the past (Kishi, 2005; Pretty and Hine, 2005). In fact, Way and Heong (2021) concluded that insecticides were not needed in most cases to control pests, and pests can be controlled by adopting physical and cultural measures that are eco-friendly. Subsequently, several agencies realized the importance of agroecological-based approaches, stating that sustainable rice intensification programs can be developed with minimal or no use of pesticides.

The Brown Plant Hopper (BPH) was a classic case that became serious because of pesticide overuse in paddy rice cultivation in Asia. The natural BPH populations were otherwise kept under control by their natural enemies and predators (Gill and Garg, 2014). It became a major threat to rice production in Asia and many parts of India, starting as a sporadic pest during 1958–1969, and later spread to other rice-growing regions by 1982 causing huge losses to rice crops (ICAR-NRRI, 2017). To combat the BPH, countries initiated IPM programs, as evident from the steps taken in Indonesia and India (Thorburn, 2015; ICAR-NRRI, 2018). The BPH offered vast experience to scientists, both positive and negative, and showed the importance and need of a better understanding of the agroecosystem, and the relevant ecological factors for developing future IPM programs. The IPM program introduced in Indonesia to control the BPH during the 1990s was supported by the innovative Farmer Field School (FFS) model – a farmer-driven agroecosystem-based crop management to grow a healthy rice crop (Thorburn, 2015). The FFS approach at that time was a significant deviation from the top-down conventional extension approach, where cooperation, co-development, and co-implementation of IPM program was emphasized, which contributed to the successful management of BPH on rice in countries where it was adopted. Thus, BPH experience became a major entry point for IPM programs in other crops. A similar initiative in the Lao PDR (2015–2017 period) was taken up, where 54 FFSs were set up involving 1,562 farmers (37% of them female). The farmers went through participatory learning about the multiple goods and services derived from paddy-based farming systems – the importance of their management through improved agronomic practices (e.g., wider plant spacing/reduced seeding rates), improved water management (alternate wetting and drying), and reduced use of chemical pesticides through the application of ecologically sound IPM and biological control (Ketelaar et al., 2018). The FFS approach gradually evolved over the last two to three decades to address a wide range of challenges and is currently implemented in over 90 countries worldwide (FAO, 2019). Other good examples of farmer participatory extension approaches were reported from India, for example, the Village Knowledge Centres (VKCs) that are information and communication technology (ICT) based knowledge exchange platforms, which encourage farmers and scientists to work together in developing IPM programs, training through participatory "plant clinics" and implementation (Swindell, 2006; Mohapatra et al., 2019b, 2021a). However, the sustainability and upscaling of these programs depend on funding, policy support, and farmer response.

Understanding the behaviors of the pest or disease alone does not help to develop a good IPM program. Farmers and scientists must get a holistic understanding of the ecology of the pest, pest-predator dynamics, multifunctional effects of the rice agroecosystem, and the benefits and ecosystem services it generates if managed well (Luka and Yusuf, 2012). This was one of the key factors emphasized by the FAO-initiated IPM. In addition, the farmers' abilities and capacity to use the information and implement the selected measures in a timely manner are equally important. According to Heeb et al. (2019), the cross-sectoral approach in IPM that aims to reduce pest-induced crop losses will also improve ecosystem services and reduce the GHG intensity per unit of food produced and make agricultural systems more resilient to climate change.

As we are aware, paddy rice or wet-land rice is a vast human-made wetland with ecological complexity. The rich diversity of arthropods in the paddy agroecosystem, both plant eating and predators, helps to keep their populations in balance, especially abundance of the former groups that provide a good source of food for the latter (Thorburn, 2015). The generalist predators (spiders, dragonfly, damselfly, water striders, etc.) in turn help to keep the rice pest populations below the economic threshold levels. Several species of host-specific parasites and parasitoids play an important role within the complex ecological web. The behavior and population dynamics of the predator-pest species within an agroecosystem is heavily influenced by the type of soil, water and crop management operations used, including the agronomic management practices, and the use of external inputs, especially synthetic pesticides and inorganic fertilizers (Heong et al., 2021). The lack of integrated measures to manage the pest with reliance on chemical pesticides alone will not be enough; rather, it leads to negative impacts on the environment and human health (Gill and Garg, 2014). Therefore, increasing awareness about a holistic IPM approach and farmers' capacity-building is important for successful adoption. In India, IPM trainings and awareness campaigns thus became a part of the nationwide IPM program, as evident from the trainings that were conducted during 1995–2021 for scientists, extension workers, and farmers through formal courses, FFSs, and exposure visits (Government of India, 2021). However, the results and outcomes of the trainings were not able to realize the expected outcomes in the field, because of the lack of systematic follow-up and other critical shortcomings. The rate of adoption of IPM technology by farmers varied and was influenced by age, education, involvement in community-based organizations, and the ability to recognize the pests and farm size (Rao et al., 2011). Other constraints linked to lack of trained extension officers on IPM, and poor farmer integration in the planning and implementation of IPM slowed down the IPM adoption.

A conceptual framework for sustainable and climate-smart integrated pest management

During the last 60 years, IPM development across several countries has illustrated how crop protection evolved over time with varied outcomes, including both success and failures (Deguine, 2021). The main objective of IPM when it was first introduced was to reduce the overuse of synthetic pesticides. During the late 1990s, the importance of sustainable intensification in agriculture gained momentum (Conway, 1999; Pretty and Bharucha, 2015). Consequently, the importance of environment and human health in IPM started to become a priority together with reducing crop damage and yield losses. More recently, with the emphasis on climate action, new IPM programs are obligated to give due consideration to the impacts they would have on climate or *vice versa*. Hence, the IPM interventions must not only be sustainable and ensure environmental health but also contribute to GHG reductions and carbon sequestration (Figure 4.1).

Figure 4.1 illustrates a pathway for transition from conventional pest management (PM) toward climate-smart and sustainable IPM. First and foremost, a good understanding of the potential of a healthy rice agroecosystem and its multifunctionality will be a necessary step in IPM. At the next level, an enabling institutional environment and policy support is essential for effective implementation and upscaling of any IPM program, including the timely and right advisory from scientific community and the genuineness of farmer participatory extension approaches used, targeted trainings, adequate investments, and necessary infrastructure.

Th choice of IPM measures must be based on the needs of the rice agroecosystem with the appropriate combination of nature based, cultural, physical, chemical, and biological measures that not only help to improve the agroecological processes (e.g., nutrient cycling, biological nitrogen fixation, soil carbon sequestration, and predator activity) but also contribute to overall sustainability of the agroecosystem (Thorburn, 2015; Ketelaar et al., 2018; Heong et al., 2021).

Habitat management in non-croplands

The main objective here is to protect the ecosystem and biodiversity around the crop lands, especially the natural enemies and pests. A proper habitat management

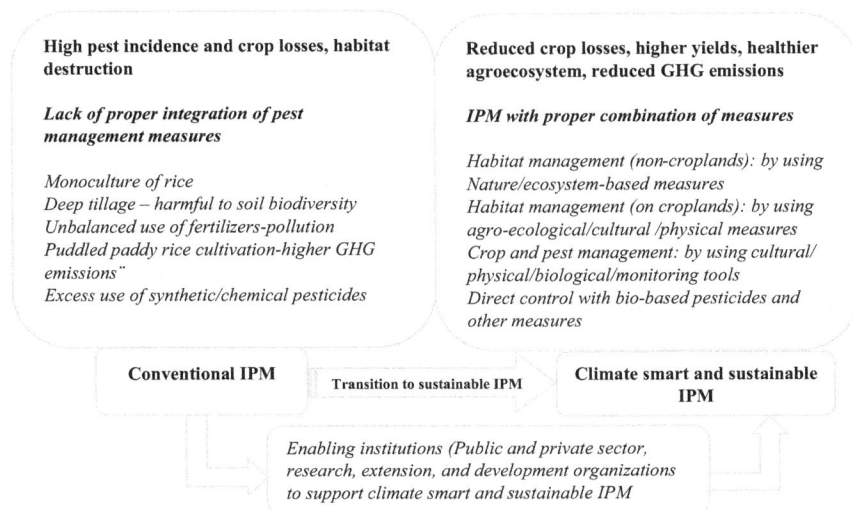

Figure 4.1 A conceptual framework showing a transition from conventional pest management to climate-smart and sustainable IPM (authors' own compilation).

helps to establish a functional link between crop and non-crop lands, improves the interactions of pest-predators, and, thereby, provides a healthy diverse habitat for both beneficial insects and alternative hosts for pests. Agroecosystem-based measures are normally recommended for habitat management that include maintaining graminaceous flora around paddy fields, planting catch crops, and growing green manure crops like Chinese milk vetch *Astragalus sinicus* after the rice harvest, which can provide shelters for native natural enemies (Huang et al., 2005). The numbers of species, individuals, and diversity index of natural arthropod enemies were found to be significantly greater in Chinese milk vetch fields than those in winter fallow fields (Yuan et al., 2010). Furthermore, the milk vetch fields provide favorable conditions for natural wintering enemies where natural enemies account for 67.9% of the total insect species. Field bunds around rice farms host a number of arthropods that can effectively regulate the rice pest populations increasing the parasitoid population within the rice fields (Gu et al., 1999; Xuetal, 2004; Zhu et al., 2015).

Habitat management within the croplands

The importance of soil biodiversity conservation in agriculture has not been appropriately considered in the past. The focus has been mostly on increasing crop productivity through agricultural intensification, whereas a sustainable intensification-based IPM approach requires the adoption of nature conservation-based measures, including conservation agriculture (CA) practices (e.g., inter-cropping, zero tillage, or minimum tillage), that benefit the diverse soil fauna and flora and improve overall soil health in the croplands. Studies have shown that microbial diversity and biological activity are higher in undisturbed soils under no tillage or minimum tillage compared to soils subject to deep plowing (Nsabimana et al., 2004; Spedding et al., 2004). Also, the abundance of mesofauna was greater where CA was practiced in comparison to compacted soils (Rohrig et al., 1998). One of the important ecosystem services generated by the soil flora and fauna is the carbon sequestration in the soils, which can be a significant contribution to the mitigation efforts in agriculture (IPCC, 2022).

Spiders constitute over 90% of the natural enemy population in rice fields, which play an important role to contain the insect pests. Simple measures such as fixing of straw bundles vertically with bamboo sticks after 15 days of rice transplanting (@20 bundles/ha), 15 cm above the water level, helped harboring 10–30 spider adults, 8–10 spider egg masses, 500–600 spiderlings, and 20–30 earwigs (Tanwar et al., 2011). Similarly planting wild sugarcane (*Saccharum spontaneum*) twigs of 4–5 feet in height and 4–5 cm in diameter after 15 days of transplanting in rice fields harbor the predators at the time of occurrence of leaf-roller, thereby suppressing the incidence of pest. This was common in parts of India, as evident from a study that showed 90% of farmers in the Benakunda village of the Ganjam district in Odisha adopted the practice of planting wild sugarcane (Mohapatra et al., 2019a). Another example is the Caseworm pest incidence in West Singhbhum district, Jharkhand state, India where farmers use fresh *parasi* leaves

(*Cleistanthus collinus*) once 5–10 days after transplanting on the insect-infested rice field using a dosage of 5 kg leaves per 100 m^2 to control the pest (Mishra et al., 2020). Similar cultural measures for habitat management within the rice fields were adopted by farmers, for example, protecting naturally occurring plants on the farm bunds, which act as a source of food for natural enemies.

Avoiding insecticide sprays in the early crop stages helps maintaining arthropod population, thereby decreasing the incidence of pest such as BPH development. In Vietnam, a multi-media campaign was used to encourage farmers to stop early season spraying, and the results showed that in provinces where the campaign was implemented, farmers reduced insecticide sprays by 53% (Heong et al., 1998). A balanced application of chemical fertilizer (major and micronutrients) can improve the utilization efficiency and rice plant vigor and enhance the resistant ability (de Kraker et al., 2000). For example, there is a positive relationship between rice resistance to pests and application of silicon in rice. Silicon can induce rice resistance or tolerance to adversities (Thripathi et al., 2014), such as stem borer, rice blast disease (Zhang et al., 2003), and the white BPH eggs laid on rice culm (Yang et al., 2014). Thus, micronutrients have multiple benefits, such as reduced pest damage and increased yields.

Crop and pest management

By using a good combination of cultural, physical, and biological measures, it is possible to manage the rice crop and relevant pests through IPM programs – for example, changing planting and harvesting timings to avoid peak pest incidence, use of pest and disease-tolerant rice varieties, use of biopesticides, rice stubble management. Rice stubbles left in the field after harvest serve as the main overwintering sites for several rice stem borers. Rice stubble management using mechanized harvesting can significantly reduce the initial population by reducing overwintering sites (Guo et al., 2013; Wu et al., 2014; Xu et al., 2015). Agroecological based measures such as intercropping and crop rotations help in suppressing pest populations, and damage to rice crop (IPCC, 2022). Some examples of rice crop and pest management are discussed later in this chapter from the authors' own field study.

Monitoring and forecasting of insect pests

Insect pest monitoring these days is becoming an important component in IPM programs as it increases the knowledge of the pest dynamics in the field that helps growers in decision-making – for instance, the intervention thresholds to counteract a given insect pest infestation, optimizing the control strategy and reduced use of chemical inputs (Mohapatra et al., 2016). Monitoring data can also be used to develop phenological models that can predict insect population outbreak, which in turn will provide additional information to improve the control techniques and optimize insecticide usage (Dent, 2000). Tools and methods such as *e-Pest Surveillance*, *Smart Light* trap, and other devices for monitoring pests are now available for improved pest monitoring and forecasting. Similarly, light traps are useful in monitoring rice stem borers and planthopper populations. The classic monitoring approach of insect pests is by placing a series of traps in infested rice fields which are monitored manually. More recently, software and image recognition algorithms have been used to support automatic trap usage to identify and/or count insect species from pictures and enable real-time and online pest monitoring (Mohapatra et al., 2019b). Farmers these days can opt for any model of light trap based on the cost and availability of their power source, i.e., direct electricity, battery, or solar energy as per the requirement/situation, with the costs varying from approximately US$12 to US$200 (Mohapatra et al., 2019b). These traps are eco-friendly, portable, easy to operate and no special skills are required for installation.

Based on the peaks of its light-trap captures, forewarning models can be developed, for example, a model for rice leaf folder was developed using the light-trap captures and weather variables such as maximum temperature, relative humidity, and sunshine hours, which could explain 99% of the variability in leaf folder light-trap peaks in a study in Punjab, India (Singh et al., 2015). Similar models were analyzed for BPH to inform farmers about the pest incidence and measures to be taken (Thakur et al., 2020). In general, pest weather relations have been analyzed through empirical models, which behave on a location-specific manner (Chander, 2010). The models such as DYMEX can forecast the impact of climate on the dynamics of the rice yellow stem borer population and generate a monthly or seasonal trend pattern, with R^2 values of 0.74 (calibration) and 0.88 (validation) (Nurhayati and Koesmaryono, 2017). In addition, mobile applications are becoming popular in agriculture, for example, the "*riceXpert*App" developed by the National Rice Research Institute (NRRI) in Cuttack is a multilingual mobile app with a "pest solution" module to identify the pest and estimate the right amount of pesticide to be used. The solution can help to create prescriptions automatically for different rice pests (Mohapatra et al., 2018, 2019b). One of the main objectives of the app is to increase the precision of pest control and avoid the use of chemical pesticides wherever possible.

Direct pest control with bio-based products

Another category is the non-toxic bio-based products that can manage pests and improve plant health without using chemicals that can harm people and benefit insect populations. The bio-based pesticides used are naturally occurring substances, such as microbes, bacteria, and plant extracts. The use of bio-based products promotes plant and soil health, while managing weeds, pests, and plant diseases in a broad range of agricultural, horticultural crops, as part of an IPM strategy. Biological insecticides are now widely accepted and commonly used in agriculture and increasingly available in the markets in India. Products derived from the neem tree (*Azadirachta indica*) were traditionally used for controlling pest in agriculture because of the presence of active ingredient Azadirachtin in neem tree (Morakchi et al., 2021). Seed treatment with bio-based products helps in reducing pest attack and damage to rice crop at a later stage. For example, rice seedlings roots-dipped in 5% neem seed kernel extract for 12 hours reduced egg laying and hatching of green leafhopper *Nephotettix virescens* (Abdul Kareem et al., 1988). Similarly, seed treatment with *Trichoderma viride* was found to be effective in controlling the rice seed-borne fungal pathogen, *Rhizoctnia solani*, and foliar-borne bacterial pathogen like *Xanthomonas oryzae* pv. *Oryzae* (Bhat et al., 2009; Tanwar et al., 2019). The studies reported that making the products easily available to farmers and regular trainings about their usage would increase adoption of these environmental friendly products.

Bacterial and microbial formulations are replacing chemical pesticides as their availability is increasing in India. In addition, new trails and initiatives with microbial formulations by private and public research agencies across India are showing promising results, which will help scientists to include the new products in IPM programs. A recent field assay of the National Rice Research Institute, India, Bt formulation showed LC_{50} ($\times 10^7$) values as 3.77, 5.29, 4.83 and 4.93, 4.42, 4.72 against third, fourth, and fifth instar larvae of *Cnaphalocrocis medinalis* respectively (Ghosh et al., 2017). The isolates of *Beauveria* and *Metarhizium* spp. were more effective in infecting 80–93.3% rice leaf folder larvae, compared to other entomopathogenic fungi that infected only 20–23.3% larvae in 96 hours (Sahoo et al., 2013).

In recent years, application of *Trichogramma* to control insect pests in rice has become popular again since it meets the needs of food, ecological, and environmental safety standards (Wang et al., 2015). The devices and technologies for releasing *Trichogramma* in rice field have been improved (Zhang et al., 2003), including the use of devices for releasing *Trichogramma* with nectar food supplement and the recent techniques for releasing *Trichogramma* by unmanned aerial vehicle in China (Li et al., 2013). The nectar food application studies carried out to improve the biological control function in a rice-based ecosystem (Gurr et al., 2016), especially on flowering plants (Zhu et al., 2012; Chen et al., 2014) and type of nectar food spray (Seagraves et al., 2011), will be useful to develop eco-friendly IPM programs in the future. A field study carried out in India by Kumar and Khan (2005) on the release of *Trichogramma japonicum* and *T. chilonis* (@ of 50,000 numbers/ha) showed a significant reduction in the tiller damage caused by yellow stem borer (*Scirpophaga incertulas*), and folded leaves by rice leaf folder (*Cnaphalocrocis medinalis*) from 50.1% to 61.3% and from 63.8% to 75.5%, respectively. Similarly, reduction in tiller damage and folded leaves varied from 78.1% to 81.6% and from 72.6% to 81.8%, respectively, when egg parasitoids were released at 100,000 numbers/ha.

There are numerous examples of non-chemical measures used and available traditionally by farmers to control pests and diseases. We need to revisit them, document the local knowledge and evidence systematically, and thereby put them together with the scientific methods of pest control while developing new IPM programs for rice and other crops. Mechanisms must be in place to acknowledge the traditional practices and incentivize farmers that adopt them, in their efforts to reduce the use of chemical pesticides.

Impacts due to climate-smart and sustainable IPM

The impacts of a well-planned climate-smart and sustainable IPM on the rice agroecosystems can be multipronged. A right combination of IPM measures can have positive impact on the environmental, socio-cultural, and economic sustainability dimensions simultaneously (Figure 4.2). Figure 4.2 illustrates some of the benefits, which are possible because of IPM. These indicators can be used to develop a baseline and subsequently evaluate the impact due to the implementation of IPM. The impacts can be direct and/or indirect, which can be assessed using quantitative and qualitative indicators. Predominantly, it was the yield and crop damage that was given importance in IPM programs, whereas impacts on the environment and human health were often ignored. It is important to show the farmers, stakeholders, and policy makers about the multifunctionality that could generate a wider support for IPM programs (Swaminathan, 2000; IAASTD, 2008; NRC, 2010). In recent years, IPM has been evolving into a more broad based integrated pest and production management, where not only pest management but multiple benefits to sustainable production are also targeted.

Studies combining selected IPM measures such as minimum tillage, agroforestry, border crops outside croplands, crop rotation with legumes, the system of rice intensification, and so on showed significant reductions in water savings, improved soil nitrogen, and soil biodiversity (e.g., abundance of earthworms), besides reduced pesticide use and increase in yield (Kartaatmadja et al., 2004; Pretty and Bharucha, 2015). The hidden health benefits of reducing or "no pesticide use" strategy are not easy to be assessed, which can be significant in most cases and the benefits can be enormous in some cases. A recent assessment of the government

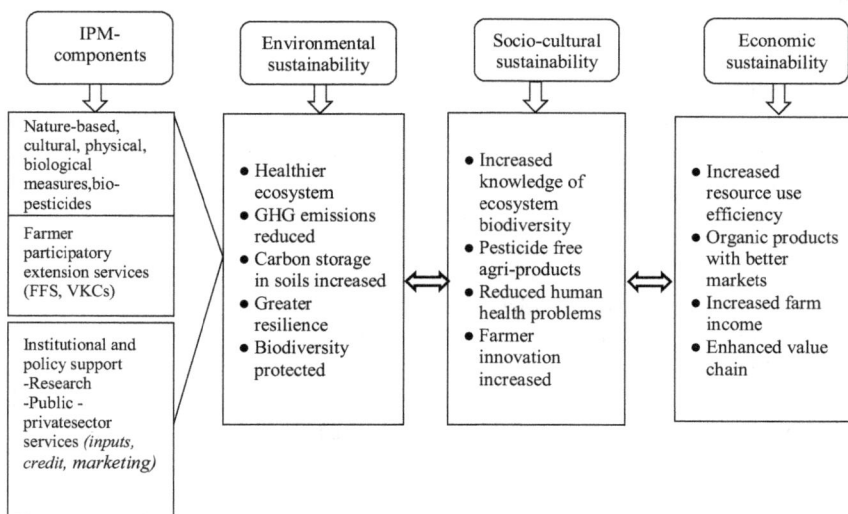

Figure 4.2 Potential impacts due to climate-smart and sustainable IPM.
Source: Authors' own compilation.

showed that IPM initiatives in India have led to an increase in crop yields from 6.72% to 40% in rice, reduction in the use of chemical pesticide by 50–100%, and an increase in the use of biopesticides from 123 million tons in 1994–1995 to 7,682 million tons by 2018–2019 (Government of India, 2021).

IPM impacts and results from Resilience and other projects in India

IPM modules involve combining several measures targeting habitat management both around and within the rice farms, in addition to the direct or specific measures included to manage the pests and diseases. In some cases, it targets a particular pest, where two or more measures are combined to manage the damage due to a single pest, whereas in others combined effort targets to manage multiple pests.

In this section, some examples of field demos of selected IPM modules in rice and their results from the ongoing Resilience project in Odisha, India (www.resilienceindia.org) which is funded by the Government of Norway are presented. As per the baseline survey conducted under the Resilience project, the major insect pests observed were rice stem borer, rice bug, and termites, and diseases such as rice blast, brown spot, and false smut that cause substantial losses to rice in the project areas were observed. About 100 lead farmers were engaged in the training and demonstrations of selected IPM modules targeting particular rice pests in the project as described below.

A simple IPM module was developed for managing brown spot disease in low land rice ecosystem, combining seedling dip in *Pseudomonas fluorescens* @ 3×10^6 cfu/ml and clipping of seedling tips and selected spraying of chlorantraniliprole 18.5 SC (soluble concentrate) @ 50g a.i. (active ingredient) per hectare at 25 days after transplanting (DAT). This module recorded lowest brown spot incidence (3.8%) compared to control treatment (12.0%).

Similarly, another IPM module for rice stem borer, combining seed treatment with carbendazim @ 1 g a.i. (@ 2 g/kg seed) and an application of chlorantraniliprole 18.5 SC @ 50g *a.i* at 25 DAT, recorded the lowest percent dead heart (2.4%) and white ear head infestation (2.0%) compared to control treatment, i.e., application of carbofuran 3G @ 3 g/m^2 (nursery application) at 7 days before uprooting of seedling and thiamethoxam 25 WG @ 25 g a.i./ha (water dispersible granule) at panicle initiation (PI) stage. Both the IPM modules – i.e., (1) chlorantraniliprole 18.5 SC at 25 DAT and *Pseudomonas fluorescens* @ 3×10^6 cfu and clipping of leaf tip at transplanting and (2) Carbendazim @ 2 g/kg seed (seed treatment) and carbofuran 3G @ 3 g/m^2 (nursery application) – registered the higher yield compared to control treatment.

Soil enzymatic activities such as dehydrogenase activity (DHA) of soil also increased by 27% in treatments with chlorantraniliprole 18.5 SC @ 50 g a.i./ha (spray) over the control treatment with thiamethoxam 25 WG @ 25 g a.i./ha. Similarly, fluorescein diacetate activity (FDA) increased by 18% due to the application of chlorantraniliprole and carbendazim (for seed treatment), and the use of *Pseudomonas fluorescens over the* application of Thiamethoxam and carbofuran (soil application), thus safer to soils than in terms of the non-target effect of these treatments on soil microbes and their activities. Chlorantraniliprole (18.5 SC) is a reduced-risk pesticide that belongs to the pyrethroid group and is considered safe to use in IPM. The overall soil microbial biomass carbon (MBC) showed an increase because of the IPM measures. These IPM measures demonstrated a positive impact on the predators as evident from the increase in predator populations such as damsel fly; ground beetle like *Paedarus* sp.; predatory bugs (*Andrallusspinidens*); and spiders such as *Pardosapseudoannulata, Tetragnatha, Xanthopimpla,* and *Tetrastichus* observed in higher numbers than the control plots.

Another IPM module demonstrated in the farmer's field at the Tangi block in Cuttack district, Odisha under the Resilience project involved a combination of pheromone traps (@ 8 traps/ha) for monitoring yellow stem borer, combined with solar powered light traps and the biological agents *Trichogramma japanicum*, an egg parasitoid at a dose of 1 gram a.i. (@ 100,000/ha released three times at an interval of seven to eight days). A need-based application of foliar spray of flubendiamide 480SC @ 30g a.i. ha^{-1} against stem borer was recommended only when the damage due to stem borer was observed. In addition, farmers were advised to place straw bundles in the field to increase the spider population and also trained to collect nymphs and adults of rice gundhi bug mechanically by using hand/sweep nets in the morning hours and reduce chemical spraying.

Results in general showed that the fields with IPM demonstrated the advantages in terms of yield ranging from 4.9 t/ha to 5.4 t/ha compared to control farmers' fields (4.8 t/ha) that used conventional PM practices. Furthermore, compared to control plots, the use of chemical insecticide was reduced by about 1.5 applications in the IPM plots. Populations of natural enemies (e.g., spider) doubled in the IPM demonstration plots. Pest populations in the IPM-treated plots were higher in the early stages, but gradually the numbers reduced than in the control fields toward the end of the cropping season because of IPM. This is in line with observations from other studies that compare IPM-treated and insecticide-free demos with conventional PM farms (Horgan et al., 2017). Higher pest numbers at early crop stages are not generally considered problematic because rice compensates for insect damage during tillering and pre-tillering stages, particularly hybrid rice varieties (Horgan et al., 2016). Farmers were generally positive about the IPM demo results that led to reduced pesticide use, increased cost effectiveness, and increased rice yields. Farmer engagement in IPM planning and trainings contributed to the adoption of IPM programs in the Resilience project villages in Cuttack.

Policy measures in support of IPM

Upscaling the IPM requires policy support and funding from the governments, and regular follow-up and monitoring of impacts. In Asia, awareness about the importance of IPM increased since the mid-1980s, thanks to the effort of international agencies such as the FAO. Take, for example, the case of India, where the central government has included IPM as one of the components in the overall Crop Production Programme since 1985 (Government of India, 1985). The IPM program received a boost with the "Strengthening and Modernization of Pest Management Approach in India in 1991–92," which was launched together with the establishment of 35 IPM centers across the country to support implementation. Further, the support for IPM was emphasized in the National Policy on Agriculture – 2000 and the National Policy for Farmers – 2007. The overall objective of IPM initiatives in India has been to increase crop production with balanced use of pesticides, minimize environmental pollution and occupational health hazards, and conserve ecosystem services. Although IPM is now a well-recognized approach in India and other Asian countries both within the government and scientific communities, in practice it still has a long way to go; we need more efforts to achieve the desired impacts. The policy and programs so far appear to be promising and are a good basis on which the new climate-smart and sustainable IPM programs must be developed.

A study by Rao and Rao (2010) in India revealed that only 3.2% of the farmers surveyed adopted IPM practices in various crops. Though IPM initiatives brought out changes in the farmers' attitude in PM and help to reduce the use of chemical pesticides, there is a need for systematic follow-up to strengthen IPM adoption. The study recommended readdressing the policies for encouraging eco-friendly options and further strengthening extension services, involving farmers as one of the top priorities. This is similar to the findings of Panda and Rathore (2017), whose study recommended that IPM programs in India must be revised and made specific to farmer and the particular agroecosystem needs.

Conclusion and way forward

Experiences so far have shown that a participatory extension-based IPM approach is the way forward to realize the full benefits, prevent yield losses in major crops such as rice, save on application of inputs, protect ecosystem, and reduce GHGs. Based on the past experiences, the future IPM programs must be prepared encompassing site-specific based, cultural and physical measures combined with biological control and/or plant origin pesticides that are environmentally friendly and climate smart. IPM technology and experiences generated by scientists working in national and international agricultural research centers must be combined with local knowledge and customized to farmer and agroecosystem needs through co-development for successful implementation. Wherever possible, new models and tools must be used to improve forecasting of pest and disease occurrence and increase precision in IPM programs.

Further research is needed to explore the full potential of natural enemies, microbial insecticides, and tolerant cultivars. Incentives for reducing use of chemical pesticides and promoting use of bio-based pesticides must be systematized. The net profits for the rice growers will increase by adopting IPM strategies properly, and thereby leading to reduced use of chemical pesticides while sustaining the yield.

Acknowledgment

The chapter has benefited from the results of the Resilience project field trials in Odisha. The authors would like to thank field enumerators and lead farmers involved in the field demos. At the same time, they acknowledge the support of the Norwegian Ministry of Foreign Affairs/The Norwegian Embassy, New Delhi for the funding and support to the Resilience project in India (2018–2023).

References

Abdul Kareem, A.A., Saxena, R.C., Palanginan, E.L., Boncodin, E.M. and Malayba, M.T. (1988) *Neem derivatives: effects on insect pests of rice and certain other crops in the Philippines (laboratory and field evaluations 1986-88I.* Paper presented at the Final Workshop of IRRI-ADB- EWC: Project on Botanical Pest Control in Rice-based Cropping Systems, 12–16 December 1988. International Rice Research Institute, Los Baños, Philippines.

Bhat, K.A., Ali, A. and Wani, A.H. (2009) 'Evaluation of biocontrol agents against *Rhizoctonia solani* Kuhn and sheath blight disease of rice under temperate ecology', *Plant Disease Resistance*, vol. 24(1), pp. 15–18.

Chander, S. (2010) 'Simulation of insect population dynamics and crop-pest interactions', pp. 74–80. In: Souvenir National Symposium on Perspectives and Challenges of Integrated Pest Management for Sustainable Agriculture, 19–21 November 2010, Dr. Y.S. Parmar University of Horticulture and Forestry, Nauni, Solan, India.

Chen, X.X., Liu, Y.Q., Ren, S.X., Zhang, F., Zhang, W.Q. and Ge, F. (2014) 'Plant-mediated support system for natural enemies of insect pests', *Chinese Bulletin of Entomology*, vol. 51(1), pp. 1–12.

Conway, J.M. (1999) 'Distinguishing contextual performance from task performance for managerial jobs', *Journal of Applied Psychology*, vol. 84(1), pp. 3–13.

De Kraker, J., Rabbinge, R., van Huis, A., van Lenteren, J.C. and Heong, K.L. (2000) 'Impact of nitrogenous-fertilization on the population dynamics and natural control of rice leaffolders (Lep.: Pyralidae), *International Journal of Pest Management*, vol. 46(3), pp. 225–235.

Deguine, J.P., Aubertot, J.N., Flor, R.J., Lescourett, F., Wyckhuys, K.A.G. and Ratnadoss, A. (2021) 'Integrated pest management: good intentions, hard realities. A review'. Available at: https://link.springer.com/article/10.1007/s13593-021-00689-w

Dent, D. (2000) 'Insect pest management', *Journal of Applied Entomology*, vol. 125, pp. 287–288.

European Commission (2016) 'Evaluation of Community Plant Health Regime'. Available at: https://ec.europa.eu/food/system/files/2016-10/ph_biosec_rules_annexes_eval_en.pdf

Food and Agriculture Organization of the United Nations (FAO) (2019) 'Farmers taking the lead-Thirty years of Farmer Field Schools'. Available at: http://www.fao.org/3/ca5131en/ca5131en.pdf

Ghosh, T.S., Chatterjee, S., Azmi, S.A., Mazumdar, A. and Dangar, T.K. (2017) 'Virulence assay and role of Bacillus thuringiensis TS110 as biocontrol agent against the larval stages of rice leaf folder *Cnaphalocrocis medinalis*', *Journal of Parasitic Diseases*, vol. 41(2), pp. 491–495.

Gill, K.H. and Garg, H. (2014) 'Pesticides: Environmental impacts and management strategies'. Available at: https://www.intechopen.com/chapters/46083

Giri, G.S., Raju, S.V.S., Mohapatra, S.D. and Mohapatra, M. (2022) 'Effect of elevated carbon dioxide on biology and morphometric parameters of yellow stem borer, *Scirpophagaincertulas* infesting rice (*Oryza sativa*)', *Journal of Agrometeorology*, vol. 24(1), pp. 77–82.

Government of India (1985) 'Integrated Pest Management'. Available at: https://pib.gov.in/newsite/PrintRelease.aspx?relid=110364

Government of India (2021) 'IPM at a glance'. Available at: http://ppqs.gov.in/divisions/integrated-pest-management/ipm-glance

Gu, D.X, Zhang, G.R., Zhang, W.Q., Qu, D.S. and Wen, R.Z. (1999) 'The reestablishment of the spider community and the relationship between spider community and its species pool in paddy fields', *Acta Arachnologica*, vol. 8(2), pp. 89–94.

Guo, R., Han, M. and Shu, F. (2013) 'Strategies and measures of green control of rice pests based on reduction of pesticides application in paddy field', *China Plant Protection*, vol. 33(10), pp. 38–41.

Gurr, G.M., Lu, Z.X., Zheng, X.S., Xu, H.X., Zhu, P.Y., Chen, G.H., Yao, X.M., Cheng, J.A, et al. (2016) 'Multi-country evidence that crop diversification promotes ecological intensification of agriculture', *Natural Plants*, vol. (2), pp. 16014.

Heeb, L., Jenner, E. and Cock, M.J.W. (2019) 'Climate-smart pest management: building resilience of farms and landscapes to changing pest threats', *Journal of Pest Science*, vol. 92, pp. 951–969.

Heong, K.L., Escalada, M.M. and Lazaro, A.A. (1995) '*Misuse of pesticides among rice farmers in Leyte, Philippines*'. In: Pingali, P.L. and Roger, P.A. (Eds), *Impact of Pesticides on Farmers' Health and the Rice Environment*. Kluwer Press: San Francisco, CA, pp. 97–108.

Heong, K.L., Monina, E., Huan, N.H. and Mai, V. (1998) 'Use of communication media in changing rice farmers' pest management in the Mekong Delta, Vietnam'. Available at: https://www.researchgate.net/scientific-contributions/Vo-Mai-2021099791

Heong, K.L., Lu, Z.X., Chien, H.V., Escalada, M., Settele, J., Zhu, Z.R. and Cheng, J.A. (2021) 'Ecological Engineering for Rice Insect Pest Management: The Need to Communicate Widely, Improve Farmers' Ecological Literacy and Policy Reforms to Sustain Adoption', *Agronomy*, vol. 11, pp. 2208. https://doi.org/10.3390/agronomy11112208

Horgan, F.G., Crisol-Martínez, E., Almazan, M.I.P., Romena, A., Ramal, A.F., Ferrater, J.B. and Bernal, C.C. (2016) 'Susceptibility and tolerance in hybrid and pure-line rice varieties to herbivore attack: biomass partitioning and resource-based compensation in response to damage', *Annals of Applied Biology*, vol. 169, pp. 200–213.

Horgan, F.G., Ramal, A.F., Villegas, J.M., Jamoralin, A., Bernal, C.C., Perez, M.O., Pasang, J.M., Naredo, A.L. and Almazan, M.L.P. (2017) 'Effects of bund crops and insecticide treatments on arthropod diversity and herbivore regulation in tropical rice fields', *Journal of Applied Entomology*, vol. 141, pp. 587–599.

Huang, D.C., Zeng, L., Liang, G.W. and Chen, Z.N. (2005) 'Population dynamics of pests and their enemies in different cultivated rice fields', *Chinese Journal of Applied Entomology*, vol. 16(11), pp. 2122–2125.

IAASTD (2008) *Agriculture at a Crossroads*. Island Press: Washington, DC.

ICAR-NRRI (2017) 'Brown Plant Hopper outbreak in Odisha'. ICAR-National Rice Research Institute, Cuttack, India.

ICAR-NRRI (2018) 'Management of Brown Plant Hopper-An old foe in new form', Available at: https://icar-nrri.in/wp-content/uploads/2018/07/2.-Management-of-Brown-plant-hopper-An-old-foe-in-new-form-2.pdf

IPCC (2022) 'Climate change 2022: Impacts, Adaptation and vulnerability', Available at: https://report.ipcc.ch/ar6wg2/pdf/IPCC_AR6_WGII_FinalDraft_FullReport.pdf

John, D.A. and Babu, G.R. (2021) 'Lessons from the Aftermaths of Green Revolution on Food System and Health', Available at: https://www.frontiersin.org/articles/10.3389/fsufs.2021.644559/full

Kartaatmadja, S., Pane, H., Wirajaswadi, L., Sembiring, H., Simatupang, S., Bachrein, S., Ismadi, D. andFagi A.M. (2004) *Optimising use of natural resources and increasing rice productivity; Conserving Soil and Water for Society: Sharing Solutions*, Proceedings of the ISCO 2004—13th International Soil Conservation Organisation Conference; 4–9 July 2004, Brisbane, Australia.

Kenmore, P.E. (1996). 'Integrated Pest Management in rice', In: Persley, G.J. (Ed.), *Biotechnology and Integrated Pest Management*, CABI International, Wallingford, pp. 76–97.

Ketelaar, J.W., Morales-Abubakar, A.L., Van Du, P., Widyastama, C., Phasouysaingam, A., Binamira, J. and Dung, N.T. (2018) 'Save and Grow: Translating policy advice into field action for sustainable intensification of rice production', In: Nagothu, U.S. (eds) *Agricultural Development and Sustainable Intensification: Technology and Policy Challenges in the Face of Climate Change.* London, Earthscan from Routledge, pp. 23–51.

Kishi, M. (2005) 'The health impacts of pesticides: What do we know?' In: Pretty, J. (Ed.), *The Pesticide Detox.* Earthscan, London, pp. 23–38.

Kumar, S. and Khan, M.A. (2005) 'Bio-efficacy of *Trichogramma* spp. against yellow stem borer and leaf folder in rice ecosystem', *Annals of Plant Protection Sciences*, vol. 13(1), pp. 97–99.

Li, D.S., Yuan, X., Zhang, B.X., Zhao, Y., Song, Z.W. and Zuo, C. (2013) 'Report of using unmanned aerial vehicle to release *Trichogramma*', *Chinese Journal of Biological Control*, vol. 29(3), pp. 455–458.

Luka, M. and Yusuf, A. (2012) 'Environmental responsibility and performance of quoted companies in Nigeria', *International Journal of Development Studies*, vol. 5(4), pp. 262–278.

Mishra, A., Singh, S.R.K. and Raut, A.A. (2020) 'Traditional Knowledge in Agriculture'. Division of Agricultural Extension, ICAR, New Delhi,pp.39.

Mohapatra, S.D. (2008) 'Participatory appraisal for biointensive IPM research in basmati rice: A case study' *Oryza*, vol. 45(2), pp. 157–177.

Mohapatra, S.D., Rath, P.C. and Jena, M. (2016) 'Sustainable management of insect pests and diseases in rice in rainfed low land ecosystem: A case study', *Indian Journal of Plant Protection,* 44(2), pp. 189–191.

Mohapatra, S.D., Nayak, A.K. and Pathak, H. (2018) 'NRRI 'riceXpert': An android app decision support tool for smart rice farming'. Available at: http://ricecongress2018.irri.org/conference program

Mohapatra, S.D., Tripathi R., Kumar A., Kar, S., Mohapatra, M., et al. (2019a) 'Eco-smart pest management in rice farming: prospects and challenges', *Oryza*, vol. 56, pp. 143–155.

Mohapatra, S.D., Nayak, A.K., Panda, B.B., Tripathi, R., et. al., (2019b) 'Integrated Pest Management Strategies for Rice', A Technical Brief, ICAR-National Rice Research Institute, Cuttack, India.

Mohapatra, S.D., Banerjee, A., Senapati, R.K., Prasanthi, G., Mohapatra, M., Nayak, P.K., Nayak, A.K. and Maiti, D. (2021a) 'Current status and future prospects in biotic stress management in rice', *Oryza*, vol. 58, pp. 168–193.

Mohapatra, S.D., Pradhan, S.S., Senapati, R.K., Prasanthi, G., Zaidi, N.W. and Nayak, A.K. (2021b) 'Rice Pest DSS in Sub' ICAR-National Rice Research Institute, Cuttack, India'. Available at: http//ricepestinformatics.in

Morakchi, S.K., Goudjil, H.M. and Sifi, K. (2021) 'Azadirachtin-Based Insecticide: Overview, Risk Assessments, and Future Directions', Available at: https://www.frontiersin.org/articles/10.3389/fagro.2021.676208/full

Nsabimana, D., Haynes, R.J. and Wallis, F.M. (2004) 'Size, activity, and catabolic diversity of the soil microbial biomass as affected by land use', *Applied Soil Ecology*, vol. 26(2), pp. 81–92.

NRC (2010) 'Towards Sustainable Agricultural Systems in the 21st Century'. Committee on Twenty-First Century Systems Agriculture, National Academies Press; Washington, DC.

NRDC (2015) 'The Story of Silent Spring'. Available at: https://www.nrdc.org/stories/story-silent-spring

Nurhayati, E. and Koesmaryono, Y. (2017) 'Predictive Modeling of Rice Yellow Stem Borer Population Dynamics under Climate Change Scenarios in Indramayu'. Available at: https://www.researchgate.net/publication/315968756_Predictive_Modeling_of_Rice_Yellow_Stem_Borer_Population_Dynamics_under_Climate_Change_Scenarios_in_Indramayu

Oerke, E.C., Dehne, H.W., Schonbeck, F. and Weber, A. (1996). Estimated losses in major food and cash crops. *Agricultural Systems*, vol. 51(4), pp. 493–495.

Oerke, E.C. (2006). Crop losses to pests. *The Journal of Agricultural Science*, vol. 144(1), pp. 31–43.

Panda, S. and Rahtore, R. (2017) 'Prospects of Integrated Pest Management in India'. Available at: https://www.researchgate.net/publication/348192129_Prospects_of_integrated_pest_management_in_India

Prasanna, B.M., Huesing, J.E., Eddy, R. and Peschke, V.M. (2018) 'Fall Armyworm in Africa: A guide for integrated pest management', Available at: at: https://reliefweb.int/sites/reliefweb.int/files/resources/FallArmyworm_IPM_Guide_forAfrica.pdf

Pretty, J. and Bharucha, Z.P. (2015) 'Integrated Pest Management for Sustainable Intensification of Agriculture in Asia and Africa', Available at: https://www.ncbi.nlm.nih.gov/pmc/articles/PMC4553536/

Pretty, J. and Hine R. (2005) 'Pesticide Use and the Environment', In: Pretty, J. (Ed.), *The Pesticide Detox*. Earthscan, London, pp. 1–22.

Rao, G.V.R. and Rao, V.R. (2010) 'Status of IPM in Indian Agriculture'. Available at: https://core.ac.uk/download/pdf/211013149.pdf

Rao, C.A.R., Rao, S.M., Srinivas, K., Patibanda, K. and Sudhakar, C. (2011) 'Adoption, Impact and Discontinuance of IPM Technologies for Pigeon Pea in South India'. Available at: https://www.researchgate.net/publication/269640363_Adoption_Impact_and_Discontinuance_of_Integrated_Pest_Management_Technologies_for_Pigeon_Pea_in_South_India

Rohrig, R., Langmaack, M., Schrader, S. and Larink, O. (1998) 'Tillage systems and soil compaction: their impact on abundance and vertical distribution of *Enchytraeidae*', *Soil Tillage Research*, vol. 46, pp. 117–127.

Sahoo, B., Behera, K.S. and Dangar, T.K. (2013) 'Selection of effective natural fungal pathogens of the rice leaf folder *Cnaphalocrocismedinalis* (Guenee) by in vitro assay', *Oryza*, vol. 50(3), pp. 278–83.

Seagraves, M.P., Kajita, Y., Weber, D.C., Obrycki, J.J. and Lundgren, J.G. (2011) 'Sugar feeding by coccinellids under field conditions: The effects of sugar sprays in soybean', *BioControl*, vol. 56(3), pp. 305–314.

Singh. P., Chander, S., Husain, M., Pal, V. and Singh, P. (2015) 'Development of a Forewarning Model to Predict Rice Leaf Folder (*Cnaphalocrocismedinalis* Guenee) Incidence in Punjab, India'. Proceedings of National Academy of Science, India, Section B, Biological Science, pp. 1–6.

Spedding, A., Hamela, C., Mehuysa, G.R. and Madramootoo, C.A. (2004) 'Soil microbial dynamics in maize-growing soil under different tillage and residue management systems', *Soil Biology and Biochemistry*, vol. 36, pp. 499–512.

Swaminathan, M.S. (2000) 'An Evergreen Revolution', *Biologist*. vol. 47, pp. 85–89.

Swindell, J. (2006) 'The Village Knowledge Centres of Pondicherry: An Indian ICT Adoption case study'. In: Leal W. (Ed.), *Innovation, Education and Communication for Sustainable Development*, Peter Lang: Frankfurt.

Tanwar, R.K., Singh, S., Singh, S.P., Kanwar, V., Kumar, R., Khokar, M.K. and Mohapatra, S.D. (2019) 'Implementing the Systems Approach in Rice Pest Management: India context', *Oryza*, vol. 56, pp. 136–142.

Tanwar, R.K., Garg, D.K., Singh, S.P., Dhandapani, A. and Omprakash, B. (2011) 'Enhancement of spider population through habitat management in rice (Oryza sativa) fields', *Indian Journal of Agricultural Sciences*, vol. 81, pp. 462–464.

Teng, P.S. (1987) 'Crop Loss Assessment and Pest Management'. St. Paul, USA: APS Press.

Thakur, A., Vashistha, U. and Sharma, S. (2020) 'Estimation of Brown plant hopper level in rice crop using Artificial Neural Network'. 8th International Conference on Reliability, Infocom Technologies and Optimization. (Trends and Future Directions) ICRITO.

Thripathi, D.K., Singh, V.P., Gangwar, S., Prasad, S.M., Maurya, J.N. and Chauhan, D.K. (2014) 'Role of silicon in enrichment of plant nutrients and protection from biotic stresses' In: Ahmad, et al. (Eds.), *Improvement of Crops in the Era of Climate Changes*. New York, USA: Springer.

Thorburn, C. (2015) 'The Rise and Demise of Integrated Pest Management in Rice in Indonesia', Available at: https://www.ncbi.nlm.nih.gov/pmc/articles/PMC4553486/doi: 10.3390/insects6020381

Wang, G.R., Han, R.P., Shen, Q., Yang, W.T., Wang, G.D., Lu, Z.X. and Zheng, X.S. (2015) 'Effects of modified fertilizer application on rice pests, diseases and yields' *China Rice*, vol. 21(6), pp. 94–97.

Wu, S.W., Peng, Z.P., Jiang, G.F., Qin, G.Q. and He, C.Y. (2014) 'Effects of harvest method and treatment of rice stubble on over wintering of *Chilosuppressalis*', *Human Agricultural Science*, vol. 1, pp. 48–50.

Xu, Y.C., Xian, D.X., Yao, Q., Yang, B.J., Luo, J., Tang, C. and Zhang, C. (2015) 'Automatic identification of rice light trapped pests based on images', *Chinese Journal of Rice Science*, vol. 29, pp. 299–304.

Yang, G.Q., Hu, W.F., Zhu, Z.F., Ge, L.Q. and Wu, J.C. (2014) 'Effects of foliar spraying of silicon and phosphorus on rice (*Oryza sativa*) plants and their resistance to the whitebacked planthopper, *Sogatellafurcifera* (Hemiptera: Delphacidae)', *Acta Entomologica Sinica*, vol. 57(8), pp. 927–934.

Yuan, W., Liu, H., Zhang, S.X. and Li, W.Y. (2010) 'Evaluation of communities of insect pests and natural enemies in organic rice fields of Changjiang Farm' *Acta Agriculturae Shanghai*, vol. 26(2), pp. 132–136.

Zhang, C.Z., Shao, C.Q., Meng, K., Li, H.L., Han, X.F. and Zhang, J.S. (2003) 'Study on rice absorbing silicon characteristics and silica fertilizer effect under salinized moist in coastal regions', *Journal ofLaiyang Agricultural College*, vol. 20(2), pp. 111–113.

Zhu, P.Y., Lu, Z.X., Gurr, G., Zheng, X.S., et al. (2012) 'Ecological functions of flowering plants on conservation of the arthropod natural enemies of insect pests in agroecosystem', *China Journal of Biological Control*, vol. 28(4), pp. 583–588.

Zhu, P.Y., Zheng, X.S., Yao, X.M., et al. (2015) 'Ecological engineering technology for enhancement of biological control capacity of egg parasitoids against rice planthoppers in paddy fields', *China Plant Protection*, vol. 35(7), pp. 27–32.

5   System of rice intensification

Empowering farmers to work with
nature to achieve productive,
resilient and climate-neutral
farming systems in rice-based
landscapes

Abha Mishra, Jan Willem Ketelaar and Max Whitten

Introduction

While the world is mid-way toward pursuing an ambitious agenda for global sustainable development to be achieved by 2030, there is an increasing realization that the associated 17 Sustainable Development Goals will not be achieved without urgent action to deal with pivotal environmental concerns, most notably the so-called Triple Challenge (biodiversity loss, climate change and pollution) (UNEP, 2021). The agricultural sector, in particular, is responsible for – as well as impacted by – this Triple Challenge. But it is also in the agriculture sector, especially the rice sector, where enormous potential and major opportunities exist for dealing with these environmental challenges.

The IPCC (2007) has reported that agriculture is one of the major sources of greenhouse gas (GHG) emissions, accounting for 12% of the GHG produced by human-made activities. Trend analysis extending activity data and GHG emissions from 1990 to 2010 indicated that agricultural emissions increased 0.7% annually between 2000 and 2010, and accelerated to 1.2% per year in the subsequent decade (IPCC, 2014). The food system alone may be responsible for one third of global gas emissions (Crippa et al., 2021).

Rice production has received major attention as a culprit but is also a potential solution from the agriculture sector for substantial GHG emission reduction (FAO, 2017). This is especially relevant for Asia, which accounts for 90% of global rice area, and 20% of the total world cropland area along with 70% of the total fresh water use for its production. Rice paddies are responsible for 5–20% of global methane emissions from anthropogenic sources (IPCC, 1996). More particularly, "business-in-as-usual scenario" will further intensify degradation of the natural resources base, such as land and water, while exacerbating GHG emissions in the region.

Therefore, GHG emission reduction from the rice sector, methane in particular, is crucially important and urgent. Recent studies on the technical potential for methane mitigation, including through application of alternate wetting and

DOI: 10.4324/9781003273172-5

drying (AWD), have shown that a 6–9 Mt/yr reduction within the rice sector is possible (UNEP and CCAC, 2021). Adoption of climate-smart practices has received considerable attention in the Asia region. In this regard, climate-smart agriculture (CSA) has emerged as one of the strategies to address the emerging needs. CSA is composed of three main pillars: (1) sustainably increasing agricultural productivity and incomes; (2) adapting and building resilience to climate change; and (3) reducing and/or removing greenhouse gas emissions, where possible (FAO, 2016).

The climate-smart agroecological-based System of Rice Intensification (SRI) defines a way forward for smallholders to grow healthy crops with less inputs (land, water, pesticides, synthetic fertilizers, labor and capital), and with greater appreciation and reliance on robust local ecosystem goods and services that suppress pests and diseases, that enhance soils, that support food security and rural livelihoods and, finally, that mitigate methane emission from the paddy system (FAO, 2014; Thakur and Uphoff, 2017; Mishra et al., 2021).

By producing stable crop yields with low external inputs, SRI also provides an opportunity for farm diversification, enhanced income generation and dietary diversity that addresses the nutritional needs of smallholders. Thereby, SRI can become the main climate change solution and support sustainable development (Thakur et al., 2021).

SRI is based on four agronomic principles:

- transplanting young and healthy seedlings (at 2.5 leaf stage) or direct sowing with relatively low seed rate;
- shallow transplanting (with minimum root disturbance) with wider spacing providing enough space and less competitive plant's micro-environment to realize the fullest potential of seedlings/seeds;
- keeping soil preferably moist, not inundated, at least during vegetative stage to allow root systems to grow larger and healthier, later maintaining shallow water level, but never creating hypoxic soil condition, thus improving plant and soil health and mitigating methane emission; and
- applying organic manure as much as possible to nurture the soil systems (feed the soil to feed the plant) (Stoop et al., 2002, 2011; Mishra et al., 2006, 2013).

All four key principles are amenable for farmers' experimentation, adaptation and adoption, as the process does not require any external physical inputs. These SRI practices – transplanting younger and single seedlings/hill with wider spacing, or direct seeding with relatively lower seed rate, giving plants more space and avoiding continuous flooding – when implemented together have, in many instances, resulted in substantial increases in yield, and reduced GHG emissions (Dill et al., 2013; Gathorne-Hrady et al., 2016; Mishra et al., 2021).

However, to have lasting impact, these ambitious goals need to be pursued in an enabling social context. Pretty et al. (2020) have argued that the political economy in past decades "prioritized unfettered individual action over the collective" thereby harming many rural institutions and "reducing sustainability and equity". Offsetting this trend, Pretty et al. note that, in recent years, changes in national and regional policy have promoted the growth of social groups whose existence supports "transitions towards policies and behaviours for global sustainability". While their study cites integrated pest management (IPM) as a catalyst in shifting the balance back from individual action to the collective, we propose that SRI is an important complement to IPM in achieving those social and economic objectives.

Thus, the chapter outlines a conceptual framework for optimizing rice productivity through responsible management of ecosystem goods and services reflective of the rich biodiversity contained in healthy rice paddy fields. Furthermore, it shares on-station research findings, coupled with participatory action research and an outreach effort at scale that has empowered thousands of farmers in their rural communities to appreciate and responsibly manage sustainable rice production. Finally, the chapter concludes with a discussion about and recommendations for a better enabling environment to allow for a real transformation toward more *sustainable* intensification of rice production at landscape, national and global levels. Such a transformation aligns well with the action called for in the UN Decade of Ecosystem Restoration (2021–2030) and is vital and urgent for the world to achieve the UN Sustainable Development Goals by 2030.

Conceptual framework for optimizing the management of ecosystem goods and services and biodiversity in healthy rice paddies while mitigating the climate change

Over the last 30 years, there have been widespread efforts to promote farmer-centered agroecological innovation using Farmer Field School (FFS) with several innovative efforts focused on the promotion of adoption of SRI (FAO, 2019; Ketelaar et al., 2020). These efforts aim to involve farmers as agent to improve paddy productivity by managing household and environmental resources efficiently. Our earlier study (Mishra et al., 2006) had conceptualized how the combination of cultural practices recommended in SRI increases the physiological efficiency of rice plants through enhanced root activity. The study also outlined many opportunities to enhance resource-use efficiency using SRI-IPM to be explored by farmers using the FFS approach.

In this chapter, we will illustrate how the paddy fields change under SRI-IPM practices. Emphasis is placed on holistic management approach of the paddy ecosystem. SRI fields that are managed without chemicals can enhance photosynthetic rate and capture more carbon and sequester more through enhanced physiological activity of roots and shoots (Mishra and Salokhe, 2010, 2011; Mishra and Uphoff, 2013). Addition of organic matter, including compost, as much as possible would restore degraded soil biodiversity, rebuild soil organic matter and restore the soil's microbiome to promote nutrients (Figure 5.1).

By following alternate wet and dry (AWD) water management practices, it is possible to improve the soil aerobic conditions, promote soil biodiversity and therefore reduce methane production and emission from paddy fields. Even under

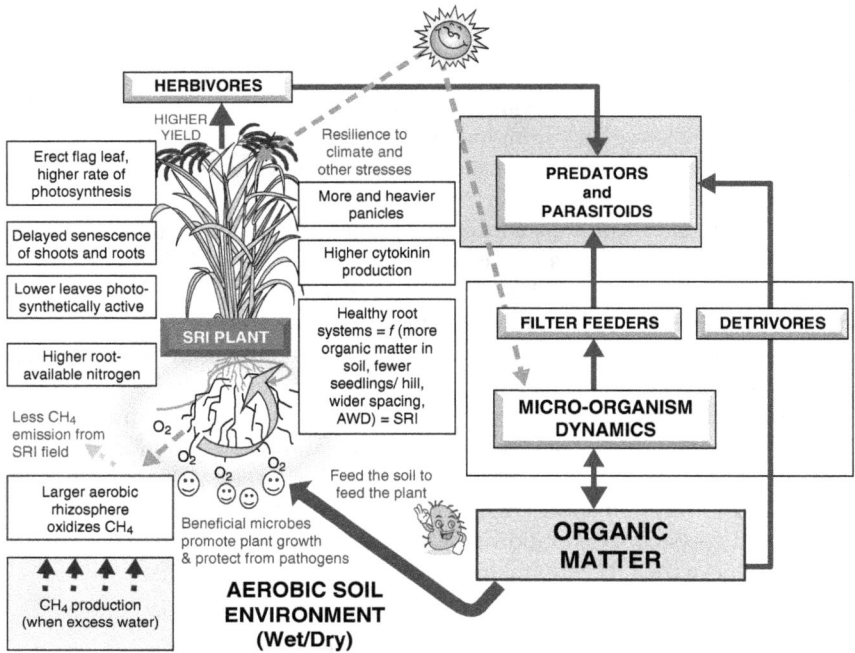

Figure 5.1 Healthy rice paddy fields under optimum management of ecosystem goods and biodiversity services facilitated by agroecological principles-led SRI-IPM practices.

shallow flooded conditions for short periods, the higher root activity would keep the rhizosphere aerobic facilitating methane oxidation and therefore less emission from the field. A healthier soil also means more CO_2 is taken out, or sequestered, from the air. Avoiding large doses of pesticides by applying IPM principles would restore and enhance the balance between pests and natural enemies as discussed in Chapter 4. In this manner, empowered farmers combine IPM and SRI to promote above and below ground natural biodiversity. A healthy soil contains a vast number of diverse microbes, which work in exchange with paddy plants growing in the soil. Rice plants absorb carbon through photosynthesis, which helps them to grow, and excess carbon is transported to the soil, where it becomes organic matter. The carbon feeds the various microbes in the soil, which in return supply the plants with the nutrients they need. A healthy soil supports a balance among all the components of the ecosystem. These sustainable intensification approaches build on sound agronomy as well as on biodiversity and ecosystem ecology for the purpose of raising crop yields and land productivity while capturing and sequestering more carbon and reducing GHG emissions.

Large-scale adoption of SRI practices can be a major game changer in terms of achieving greater land productivity while addressing key environmental challenges, including climate change. Below findings from on-station research as well as engagement at the regional level in the Mekong River Basin region provide some useful insights on the landscape-level effects: how SRI can sustainably intensify rice production and mitigate GHG emission from the paddy fields.

On-station research findings on the effects of SRI practices on root architecture and its impact on methane reduction

In the above section, we presented a model of a paddy production system that optimizes land productivity as well as the management of ecosystem goods and services and biodiversity contained therein while capturing and sequestering more carbon and reducing GHG emissions. Nurturing healthy root systems along with healthy soil is one of the priorities to realize the benefits that we have outlined in this model and the above explanatory section. In this section, we explore how healthy root systems, i.e., enhanced root growth through morphological and physiological plastic response observed under SRI practices, are linked to methane emission mitigation.

Root growth can be altered through various mechanisms (by genetic and by micro-environmental manipulation). But given climate change concerns, methane emission from paddy fields, as well as water constraints facing the rice sector in many countries, the most important crop management practice, which attracts major attention, from farmers, researchers as well as policy makers, is the cessation of continuous flooding, either through intermittent irrigation or by keeping soil moist but preferably not continuously inundated. The intermittent irrigation in rice or AWD is not something new. In fact, this has been supported earlier in some rice-producing countries in an attempt to reduce the volume of irrigation water used. The pros and cons of intermittent irrigation have been well reviewed and referenced in our previous published study (Mishra and Salokhe, 2010). It has been suggested that intermittent irrigation or AWD can reduce water use in rice cultivation by 15–25% without affecting yields, and can lower methane emissions by 30–70% (FAO, 2013). In addition, further reduction in methane emissions can be realized through plastic response of shoot and root, the latter resulting into a more aerobic rhizosphere (Mishra and Salokhe, 2011).

Research findings on the **plastic response**[1] of rice plants that resulted due to change in water regimes documented that rice root morphology and physiology and consequently rice shoot growth are significantly affected by variations in soil water conditions (Mishra and Salokhe, 2010, 2011). Root architecture (root length density) and roots' oxidizing activity rate are important factors influencing higher yields. Such responses are quite plastic in nature and vary considerably with varying water regimes and with varying soil microbial populations (Mishra and Uphoff, 2013). Modifying water management to take advantage of plants' inherent plasticity of morphological and physiological response can be one of the adaptive strategies for achieving higher yield under reduced water condition along with mitigation of methane production from rice fields (Mishra, 2019a).

The earlier findings have also shown that intermittent irrigation during vegetative growth stage along with sparse planting density of younger seedlings, i.e., single seedling transplant/hill with 20 × 20 cm spacing, increased root and shoot growth. These practices induce different levels of physiological activity, resulting into higher yields and higher dry matter production (Mishra and Salokhe, 2010). Building on the earlier findings we estimated rate of oxygen release using the Kirk (2003) root' model calculation to understand the implication of morphological and physiological plastic response on nutrient uptake and methane emissions from rice fields.

Kirk (2003) indicated that for an aerobic rhizosphere, spacing is critical along with number of primary roots per plant. Keeping the standard estimation typically observed under average paddy fields in Asia,[2] we calculated the rate of release of oxygen under different planting densities and water regimes. The calculation revealed that at flowering, the rate of release of oxygen per cm^2 was higher under P2 planting density (single seedlings/hill with 20 × 20 cm spacing) in both IV-F (intermittent irrigation during vegetative stage) and CF (continuously flooded) water regimes whereas it was drastically reduced at 20 days after flowering (DAF) under flooded condition. The effect of spacing along with single seedlings/hill had the maximum effect compared to other treatments.

Typically, maximum rates of N uptake by rice crops are 40 pmol cm^{-2} (soil surface) s^{-1}. Therefore, if half the O_2 released from the roots was used to nitrify NH_4+ in the rhizosphere (($NH_4+ + 2O_2 \rightarrow NO_3- + 2H+ + H_2O$) and half the NO_3- produced was recovered by the roots, an O_2 release of 160 pmol cm^{-2} would be sufficient to nitrify half the nitrogen absorbed by the roots. This is one of the prerequisites, i.e., 50% N uptake should be in the form of NO_3 to achieve the higher yield under any field condition (Wang and Below, 1996; Briones et al., 2003).

Furthermore, it is known that up to 90% of the CH_4 emitted in rice paddies is released through rice plant transport (Conrad, 2007), while between 19% and 90% of the CH_4 produced is oxidized, with up to 75% of the CH_4 oxidation taking place in the rhizosphere (Frenzel, 2000). Accordingly, strategies to lower net CH_4 emission from rice fields include reducing CH_4 production, increasing CH_4 oxidation and lowering CH_4 transport through the plant. Among the CH_4 emission mitigation strategies that do not compromise rice productivity, the introduction of drainage periods during the crop cycle appears to be the most efficient (Neue, 1993). Thus, it has been estimated that intermittent drainage periods by applying intermittent irrigation in poorly drained rice fields could reduce to 10% of the agricultural CH_4 emissions (Kern et al., 1997).

It is expected that the higher root activity rate (Mishra and Salokhe, 2010) along with higher release of oxygen for a longer duration (Figure 5.2), as appeared in our studies, should further enhance CH_4 oxidation in the rhizosphere because of the prolonged oxygenated rhizosphere.

This benefit will be relatively higher under intermittent irrigation water regimes but even under flooded condition (though relatively less) at flowering stage if intra-hill competition is minimized to keep rice rhizosphere aerobic (Figure 5.2). Optimizing planting density per unit area through SRI not only enhanced root activity but also increased oxygen release under single seedlings/hill planting option. Thus, maintaining an aerobic rhizosphere for longer duration would not

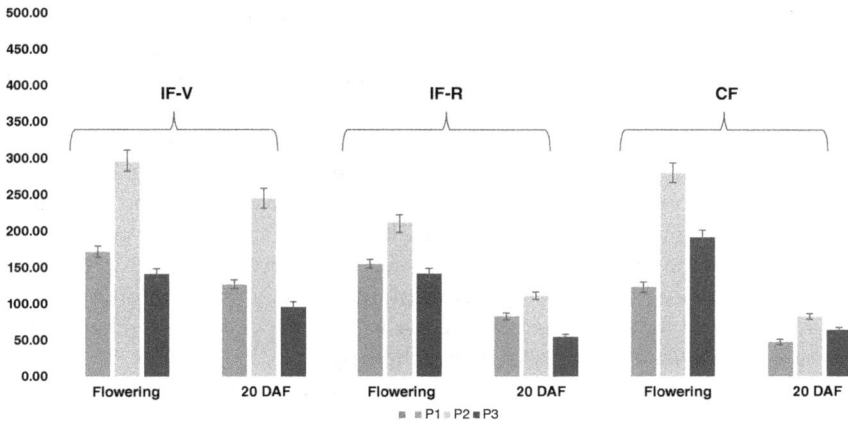

Figure 5.2 Effects of planting density.

only facilitate higher uptake of nitrogen in the form of nitrate and ammonium for higher biomass production but also support methane oxidation in the rice rhizosphere by 75–90%. Thus, there will be negligible methane emission from the rice fields. Indeed, optimization in spacing and water management is needed in order to enhance the oxygen release in rhizosphere that benefits plants, soil microbes and environment without making it burden for plants as releasing oxygen in rhizosphere is an energy-consuming affair for plant.

In addition, intermittent irrigation will reduce the aerenchyma formation rate (Mishra et al., 2006). Since the aerenchyma acts as a channel for oxygen transport from the atmosphere to the roots and CH_4 transport from the site of production to the atmosphere, reduced aerenchyma formation will lead to lowering CH_4 transport through the plant. These benefits become more relevant in the prospective scenario where rice production needs to be increased with both reduced water applications and reduced "climate-forcing" practices.

Scaling up through farmers' participatory action research (FPAR) for transition toward agroecological-based SRI methods

As discussed above and reported by many, SRI is considered as one of the best currently available agroecological methodologies for sustainable intensification of rice-based farming systems that involve farmers as agents to improve productivity by managing environmental resources efficiently. We piloted several plot scale efforts in the Lower Mekong River Basin countries to explore the usefulness of SRI and IPM practices using the FFS approach for having farmers learn about sustainable paddy production (Mishra and Kumar, 2009; Mishra et al., 2013). The outcome of these collaborative initiatives gave impetus for scaling-up efforts to the regional level. To learn more about the SRI's usefulness for fueling innovation at grass-root level involving smallholder farmers as a main agent, a regional collaborative project, funded by the European Union, was implemented in rainfed areas of the Lower Mekong River Basin (LMB) countries (Cambodia, Laos, Vietnam and Thailand) involving more than 15,000 smallholder farmers directly (and 30,000 indirectly), researchers, extension personnel and development professionals, together with staff of relevant government ministries (www.sri-lmb.ait.asia/).

The six-year-long project's objectives were to fuel local innovation to produce healthier, more profitable rice crops with less energy and a lower carbon footprint by using the SRI method under rainfed conditions.[3] The idea was also to learn and advance knowledge on technical, institutional and organizational innovation needed for shift from ready-to-use to tailor-made solutions that address location-specific issues along with global challenges.

Keeping this in mind, local, national and regional innovation platforms were designed to systematize engagement and strengthen communication for fueling innovation. This was a network building effort that was initiated by the project and was expected to continue as a common meeting point at all levels. These platforms facilitated policy dialogues on food security, research for development, marketing improvements and extension capacity for the rainfed LMB region. More than 15 institutions (academic, research and development) were involved in the six-year-long farmers' participatory action research (FPAR) field trials located in the 33 rainfed districts of 11 provinces in the LMB countries (Cambodia, Laos, Thailand and Vietnam). SRI's agronomic principles were used as "entry points" for such engagement-led transition. The number of farmer-participatory experiments conducted was more than 1,052 at >500 sites across the LMB region. As a part of this FPAR intervention, the common issues and interests expressed by farmers producing under rainfed conditions in all four countries were to achieve higher yield with reduced costs of production by reducing input use for cost saving and for making rice cultivation more efficient and profitable.

Using results of baseline survey, including information generated through various group-dynamic tools such as sub-group discussion, visual tools and brainstorming sessions, a range of experimental options were selected for each of the target areas that revolved around the integration of a few SRI principles with existing conventional practices to be applied on a learning plot for location-specific adaptation. As part of these field experiments, farmers were also encouraged to apply the full set of SRI principles on a demonstration plot which served as a "test site" to show the full potential of SRI methods at smallholder farmers' field level. For comparison purposes, the practices that were applied were categorized into (1) **Baseline (indicated as CP)**, (2) **Farmer's practices** (FP), (3) **SRI-demonstration** (SRI-D) and (4) **SRI-transition** (SRI-T).

Baseline – the existing conventional management practices (CP) generally followed in the target area as identified through the baseline survey prior to action research set-up.

Farmer's practices (FP) – the existing management practices generally followed in the target areas and set up by farmers as FP plots for comparisons purpose during the action research field experiment set-up.

SRI-demonstration (SRI-D or SRI) – where the full set of SRI practices was applied.

SRI-transition (SRI-T) – where a few principles of SRI were applied in combination with modified or existing conventional practices by farmers. The word "transition" was used because the practices are generally transitioning toward SRI with different degrees of SRI adoption and types. These practices do not fall in either category of SRI or FP. Instead, these practices were modified by farmers, improved and considered better than FP. These plots were also termed as "learning plots". Details of the experiment along with the specific details of the SRI-D or SRI, SRI-T and CP alternatives can be seen in Mishra et al. (2021).

Aligned with the Farmer Field School (FFS) interventions, FPAR structure and research/outreach implementation design was established (Figure 5.3). At some places, the structure was adapted based on the existing local government extension department's program implementation structure and also according to the farmer's needs and requirements. The design involved 50% women (at least) and 10% landless to have an inclusive intervention.

This structure facilitated the systematic introduction of SRI/IPM/FFS approaches for the development of knowledge-intensive and location-specific technologies by bringing farmers, researchers, trainers and other stakeholders

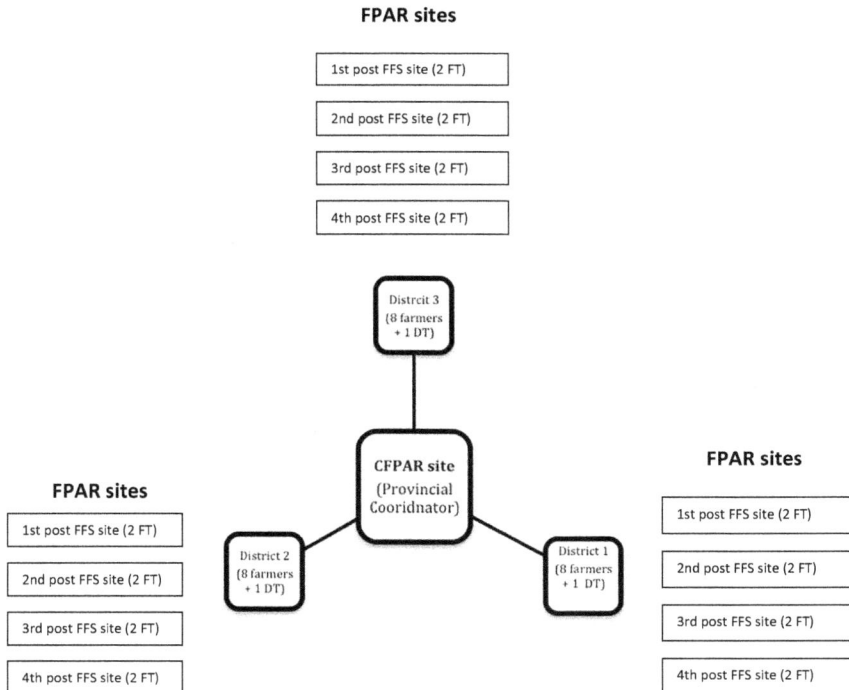

Figure 5.3 Structural diagram of CFPAR and FPAR in one province.

together, and by fueling their innovative capacity. Apart from these tangible and quantifiable direct benefits to the target groups of farmers, locally developed technologies for rice and other crops could take a horizontal spread pathway and reach other farmers in neighboring communities (approx. 50,000 farmers, based on past FFS experience in the region) through field days. Through this learning-centered approach, we also refined the curricula options for women and landless in order to capitalize on the opportunity that the action presented for furthering the leadership and empowerment of women, especially in household decision-making and income generation activities. The process of engagement led to the development of informal farmers' groups and networks in all four countries.

The FFS interventions also facilitated systematic data collection. The data were compiled at provincial, at country and, finally, at regional level using online project database (see User's guideline for online database: http://sri-lmb.ait.asia/downloads/User's%20guide%20on%20online%20database%20for%20SRI-LMB.pdf).

Furthermore, the structure helped in creating a way forward for participatory policy and program development for ensuring better market access, price and returns, also as a step toward NDCs contribution under the Paris Agreement along with achieving its SDGs. The research conducted on the policy environment, and the institutional responses to the adaptation, revealed that the adaptation and adoption of agroecological practices like SRI in the region need to be further strengthened realizing that the macroeconomic situation across LMB countries is at different stages of development, and still evolving (done by Oxfam and compiled in Mishra, 2019b).

Nevertheless, the results of this collaborative engagements showed that SRI principle-led practices helped to improve conditions in rainfed areas across the LMB region in numerous ways: average rice yield increased by 52%, and net on-farm economic returns were raised by 70% because of lower production costs. Labor productivity was increased by 64%, water productivity by 61% and the efficiency of mineral fertilizer use rose by 163%. The total energy input required for farming operations was decreased by 34% per hectare (Mishra et al., 2021). The data also showed that per hectare emissions of GHG were significantly reduced, by 14% with irrigated rice production, and by 17% from a lower level in rainfed cropping (Mishra et al., 2021, Table 5.1) due to less input usages. In terms of average reduction in GHG emissions from SRI fields in the three countries (Thailand, Vietnam and Laos), it was 25% from irrigated and 30% from rainfed systems.

Therefore, the eco-efficiency (*Ecoefficiency (USD/tCO$_2$ eq/ha/year) = Net income (USD)/Total GHG emissions (tCO$_2$ eq/ha/year)*) was also increased under SRI and SRI-T practices compared to the baseline. Interestingly, when the eco-efficiency of rainfed and irrigated production systems was compared, it was found that rainfed production was more eco-efficient compared to irrigated systems (Figure 5.4). These findings were based on the general bio-physical properties of the production environment where trials were conducted and on the cropping pattern followed in those areas (for details see Mishra et al., 2021).

If we look across the four countries, 64% area is rainfed (74.1% in Thailand, 83.7% in Cambodia, 41.1% in Vietnam and 81% in Laos), and only 36%

Table 5.1 Greenhouse gas (GHG) emission (tCO$_2$eq/ha) from SRI and conventional fields

Countries	Irrigated			Rainfed		
	SRI	Baseline	% change with respect to baseline	SRI	Baseline	% change with respect to baseline
Thailand	1.86	2.52	−26	1.42	2.07	−31
Vietnam	2.35	2.92	−20	1.9	2.48	−23
Laos	1.17	1.74	−33	0.73	1.3	−44
Cambodia	2.20	1.54	42	1.76	1.09	61
Regional (av. of four countries)	1.89	2.18	−13	1.45	1.74	−16.67
Regional (av. of three countries: Thailand, Vietnam and Laos)	1.79	2.39	−25.1	1.35	1.95	30.76

Source: Based on authors' own data from field.

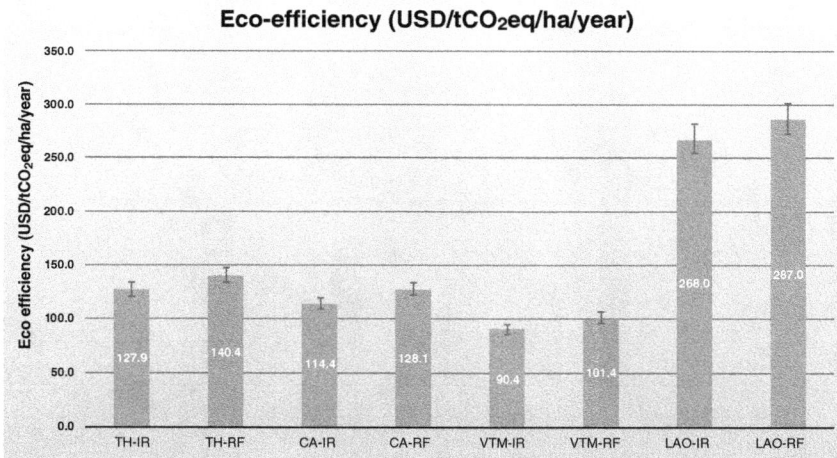

Figure 5.4 Average eco-efficiency of irrigated and rainfed production systems with SRI practices in LMB countries.

is irrigated in the region. Using the data from our research findings to calculate the GHG emissions from current conventional practice for the LMB region as a whole, it was estimated to be 6.41 million tCO$_2$eq from rainfed and 5.18 million tCO$_2$eq from the irrigated regions of the four countries. With adoption of SRI-D methods, the GHG emission will be 5.13 million tCO$_2$eq from rainfed and 4.11 million tCO$_2$eq from irrigated rice areas, which is an overall reduction of 20% (Figure 5.5).

GHG Emissions Million t CO₂ eq. for COUNTRY IRRIGATED AREA

GHG Emissions Million t CO₂ eq. for COUNTRY RAINFED AREA

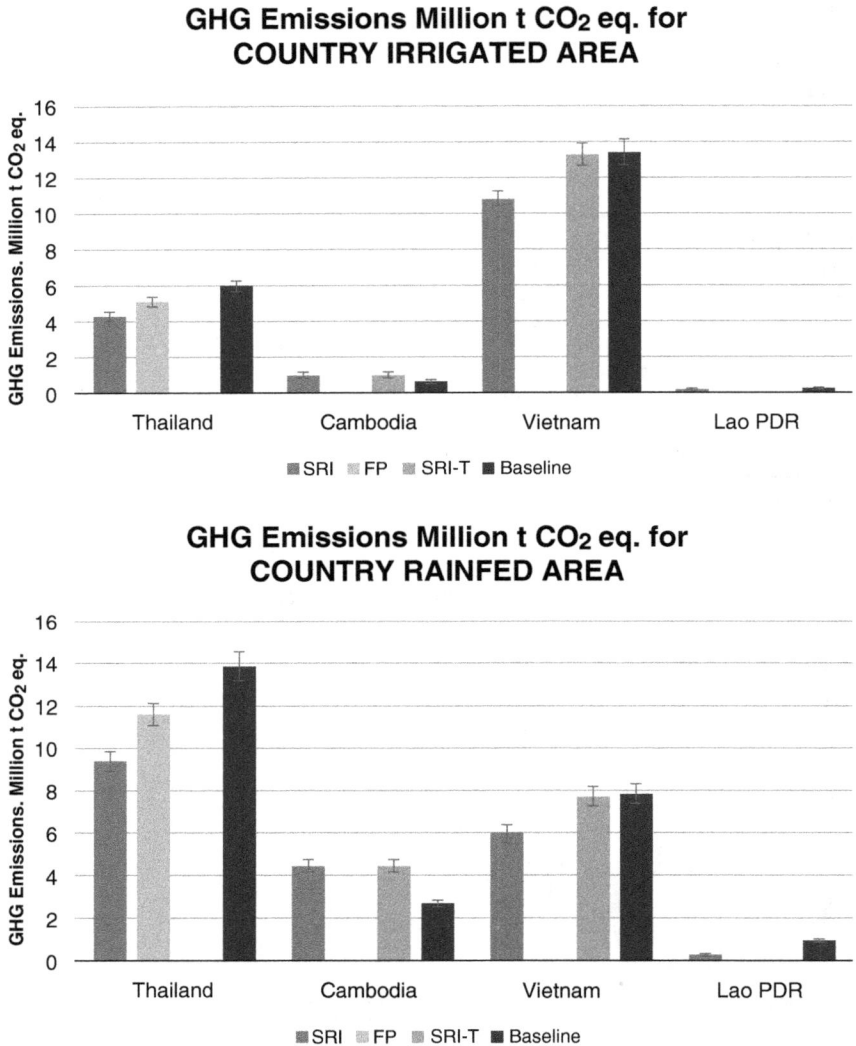

Figure 5.5 GHG emission estimation in SRI, SRI-T and FP at country and regional levels in irrigated and rainfed systems of the LMB region.

These figures show that the absolute gain for GHG reduction could be more in the rainfed parts of the region compared to the irrigated ones. The rainfed production systems have not received much policy attention but there is an opportunity to investigate "underexplored" production systems that can provide multiple wins: addressing food security and nutrition of smallholders, poverty alleviation, natural resource conservation, restoring and nurturing ecosystems and creating circular economy and supporting socially just equitable development.

Discussion: Opportunities, challenges and recommendations

Rice farming systems of the future must produce more grain while minimizing environmental impact to achieve sustainable food security goals. Higher yields and increased resource-use efficiencies are not necessarily conflicting goals and recent publications indicate that there remains tremendous scope for productivity gains and higher resource-use efficiencies in most rice production landscapes (FAO, 2014, 2016). In particular, pesticide use and nitrogen balance per unit of production are disproportionately higher in a number of cropping systems in Southeast Asia and South Asia. Our results suggest that greater adoption of agroecological approaches (e.g., SRI/IPM) for sustainable rice intensification would increase yield with increased resource-use efficiency and reduced carbon footprints.

SRI provides opportunities for climate-smart paddy production and has opportunities to become a valuable climate solution. SRI agronomic practices increase deep root system expansion and thereby allocating more carbon to the deeper soil layer. Further, with higher photosynthetic rate and hence higher capture of CO_2 from the atmosphere for longer duration (Mishra and Salokhe, 2010) along with the application of organic manure and/or compost (as much as possible), the carbon content in soil will be increased. Therefore, soils can also store more water (either rainwater or irrigation), and this is particularly important for rainfed production systems. Given that the water stored in the soil serves as the source for 90% of the world's agriculture production (FAO, 2021), it is therefore evident that soil with higher carbon content would be more productive with clear implication for achieving higher levels of global food security.

The multiple benefits of these practices especially with regard to climate change and water uses would be more appreciated when production practices recognize and value the total virtual water usages (rainfed plus irrigated), along with other environmental footprints, such as conservation of ecosystems services. The World Economic Forum's global risks survey has identified water scarcity as one of the top five global risks affecting people's well-being (WEF, 2021). And it is known that for each 1% increase in soil organic matter, soil can store an additional 20,000 gallons of water/acre (about 0.4 ha). Indeed, the importance of increasing soil organic carbon has been stressed in recent years, presenting an opportunity to help addressing water scarcity issues along with meeting the targets of the Paris Agreement.

In addition, certain SRI practices such as intermittent irrigation and fewer rice seedlings/hill transplanted with wider spacing create conditions for a larger and longer lasting aerobic rhizosphere. These conditions enable more CH_4 oxidation in the rhizosphere, reducing methane emissions from rice fields (Mishra, 2019a,b). In practice, SRI encourages location-specific adaptation for such gains to materialize. This requires active participation between farmers, researchers and other stakeholders to ensure a successful and evidence-based shift toward a more sustainable intensification of rice-based farming systems and landscapes (Ketelaar et al., 2020).

Our regional collaborative multi-stakeholder's intervention confirmed that the SRI-IPM-FFS empowerment philosophy can contribute to poverty reduction while making smallholder farming more attractive, profitable and smart. Smallholders' rice farmers can feed themselves and help feed the rest of the world while minimizing the environmental costs of agricultural intensification. Such efforts can be extended to farming systems of many scales provided that farms' heterogeneity is taken into consideration while implementing the action. This was encouraged through the flexibility that is part of the SRI approach.

The results also showed the contribution of rainfed production systems in terms of GHG mitigation aligned with sustainable development goals and the Paris Agreement. Currently, rainfed rice accounts for 33% of the total rice production area in the world. Because of relatively low yields (2–3 t/ha compared to the global average of 4.5t/ha) (Hayashi et al., 2018), rainfed rice production landscapes provide only 19% of world rice production. Therefore, improving productivity in rainfed rice with active community participation is imperative not only to lessen the burden on irrigated production systems but also to enhance efficient use of green water and other natural resources for food supply for local as well as global population. The effort will not just address poverty, food insecurity, rural development and support ecosystem benefits. It will also alleviate the pressure on irrigated rice production, which withdraws 70% of fresh water and is blamed for the bulk of CH_4 emissions.

The Southeast Asian regional policy, presented in the ASEAN Integrated Food Security Framework and the Strategic Plan of Action on Food Security, has highlighted a strategy to attain long-term food security and improve the livelihoods of farmers in the region (ASEAN, 2015). Particularly in the context of climate change, the Strategic Plan of Action on Food Security for 2015–2020 included as one of its "strategic thrusts" the introduction of climate-smart agriculture in the ASEAN member states. Accordingly, the plan recommended pilot testing of integrated technologies and practices, such as SRI and CA (Conservation Agriculture).

However, there is as of yet little evidence for the implementation at scale of such recommendations, not just for SRI but for any agroecological practices that demand knowledge and skill-intensive capacity building. Without substantial investments in farmer education and empowerment, it is unlikely that greater adoption rates of such knowledge-intensive agroecological practices will ever be realized. In tandem with such investments and other enabling conditions, a real transformation of the rice sector will require major changes in policies and regulations as well as the removal of perverse subsidies, including on chemical fertilizers and pesticides (FAO, 2016; Ketelaar et al., 2018).

That said, Vietnam is showing the ASEAN region that national policy support combined with enabling conditions, including quality capacity building interventions (e.g., Farmers Field Schools) for scaling out agroecological rice intensification (e.g., SRI, IPM), can lead to major advances in sustainable development. Some 4 million rural households in Vietnam currently apply SRI each production season with the government investing annually in farmer training and outreach activities (MARD, 2016). In 2020, the Vietnamese government declared SRI a climate-policy "breakthrough" to increase agricultural production while reducing methane emissions from rice paddies. Other ASEAN countries could do well to follow this great example and put similar policies and actions in place.

Greater investments in the rice sector as well as adoption of sustainability standards, such as the Sustainable Rice Platform Standard,[4] should also facilitate a transformation in the rice sector toward greater sustainability, including mitigation of GHG emissions.[5] While international and domestic markets in the ASEAN region are increasingly concerned about sustainability issues, adoption of the SRP Standard can help de-risk supply chains and therefore also entice the private sector to invest. SRI fits perfectly well within the scope of the good practices promoted by the SRP Standard and greater adoption of SRI in the rice sector can assist countries deliver on key SDGs, including Zero Hunger (SDG2) and Climate Action (SDG12) (Thakur et al., 2021).

Apart from social barriers, limited infrastructure, limited market connectivity, conflicting connection between policies and actions and incompatibility, there are limited innovative approaches available and/or being generated to shift from ready-to-use technological solutions to tailor-made solutions for a genuine transition toward greater sustainability in the rice sector. In short, there is considerable focus on the rhetoric of the "what" and "why" rather than local farmer engagement on the "how" for co-generated solutions that work for the millions of rice producing smallholder farmers around the world.

The effort for redesigning agroecology-based sustainable paddy systems requires technical, institutional and organizational change through innovation in agricultural practices and substantial investments, including in capacity building of smallholder farmers. Equally, multi-actor agricultural innovation platforms along with the value chains should be based on demand-driven survey and analysis, as we did in our intervention. Extension and advisory services should become more flexible, user-driven and focused on local problems. Knowledge-intensive agroecological practices, such as those promoted in SRI and IPM, require substantial investments in capacity building of extension systems and their main beneficiaries, smallholder farmers (Swanson and Rajalathi, 2010).

Such transformed extension and advisory systems should also assist on utilization of different dimensions of interaction and learning processes such as farmer to farmer, market actors to farmers, input suppliers to farmers, processors to farmers, public and private extension and advisory services to farmers. And on top of that, such efforts should generate returns on four capitals: social capital (creating jobs, education and business); natural capital (restoring biodiversity, soil, water quality and carbon); financial capital (realizing long-term sustainable profit); and inspiration (being a sort of emotional or psychological capital) by giving a sense of purpose, value and relevance to people.

The current modern industrialized production system relies mainly on the highest returns on investments, but without sufficient regard to the sustainability and other negative externalities. The SRI-like approach requires integration of ecology with economics, and values empowerment and sustainability.

To date, IPM is often recognized as an ideal entry point to farmer empowerment and a re-invigoration of social capital in sustainable food production, e.g.,

Pretty et al. (2020). In this chapter, we would contend that SRI, to the extent that it embraces IPM, is perhaps a bolder platform for optimizing social capital for securing sustainable rice production. Additionally, as noted by Uphoff and others, both IPM and SRI principles extend well beyond rice and have relevance for other crops, under the label "System of Crop Intensification (SCI)", and therefore have wider relevance for sustainable food security.

Thinking more broadly, the COVID-19 pandemic has revealed how closely, human, animal and ecological health and production systems are linked. The crisis has revealed the socio-ecological fragility of current industrialized-globalized food systems and the effects on farming and food supply. Indeed, a transition to more socially just, ecologically resilient, localized food systems that requires less energy and has potential to become climate-smart solution is urgently needed. With the changing pandemic-led global environment, the year 2021 has seen the convergence of national as well as global policy goals. For example, the India delegation at the COP26 highlighted the need for global support for local adaptation for climate change amelioration. The UNEP and FAO jointly declared a formal call for the protection and revival of millions of hectares of ecosystems all around the world for benefit of people and nature. The UN Decade on Ecosystem Restoration aims to prevent, halt and reverse ecosystems' degradation. Implementing the UN message can help to end poverty, combat climate change and prevent further biodiversity loss. However, this would require adequate investments to support farmers implementing the climate mitigation measures. Empowered farmers, as custodians and managers of their local biodiversity (Whitten and Settle, 1998), are critical to realizing the UN's aspirations. Integrating SRI-IPM with the FFS empowerment approach provides a compelling example of a way forward.

Notes

1 The ability to change phenotypically/physiologically and adapt in response to variations in the environment depending on the epigenetic processes, often termed as phenotypic plasticity.
2 We used FO_2 A_R (where FO_2 = flux of oxygen across root surface, and A_R = surface area of roots capable of absorption) = 0.2 nmol s^{-1} (which is standard rate under flooded condition).
3 Managing water in a rainfed conditions through prescriptive intermittent irrigation method was a challenge and therefore not recommended in our intervention. Although there was broader realization to avoid continuously deep flooded soil condition, efforts were made to avoid continuously flooded soil condition, where needed, to keep the root system healthy. In fact, rainfed environment is naturally intermittent flooded and indeed not continuously flooded.
4 Sustainable Rice Platform – Feed the world. Sustainably.
5 https://foodplanetprize.org/entry/reducing-methane-emissions-from-rice/

References

ASEAN (2015) 'ASEAN Integrated Food Security (AIFS) Framework and Strategic Plan of Action on Food Security in the ASEAN Region (SPA-FS), 2015–2020'. Available at: https://www.asean-agrifood.org/?wpfb_dl=58

Briones, A.M., Okabe, S., Umemiya, Y., Ramsing, N. B., Reichardt, W. and Okuyama, H. (2003) 'Ammonium-oxidizing bacteria on root biofilms and their possible contribution to N use efficiency of different rice cultivars', *Plant and Soil*, vol. 250, pp. 335–348.

Conrad, R. (2007) 'Microbial ecology of methanogens and methanotrophs', *Advance Agronomy*, vol. 96, pp. 1–63.

Crippa, M., Solazzo, E. and Monforti, F. (2021) 'Food systems are responsible for a third of global anthropogenic GHG emissions', *Nature Food*, vol. 2(3), pp. 1–12.

Dill, J., Deichert, G. and Le, T.N.T. (2013) 'Promoting the system of rice intensification: Lessons learned from the Tra Vinh Province. GIZ and IFAD'. Available at: https://wocatpedia.net/images/f/f2/Giz2013-0503en-rice-lessons-learned-vietnam.pdf

Food and Agriculture Organization (FAO) (2014) 'The multiple goods and services of Asian rice production systems', Rome. ISBN 978–92–5–108447-2'. Available at: https://www.fao.org/3/i3878e/i3878e.pdf

Food and Agriculture Organization (FAO) (2017) 'The future of food and agriculture: Trends and challenges'. Available at: https://www.fao.org/3/i6583e/i6583e.pdf

Food and Agriculture Organization (FAO) (2019) 'Farmers taking the lead: thirty years of Farmers Field Schools'. Available at: Farmers taking the lead: thirty years of farmer field schools (fao.org)

Food and Agriculture Organization (FAO) (2021) 'Key messages, global symposium on soil organic carbon'. Available at: https://www.fao.org/about/meetings/soil-organic-carbon-symposium/key-messages/en/

Food and Agriculture Organization (FAO) (2021) 'Bringing climate change adaptation into Farmers Field Schools: A global guidance note for facilitators'. Available at: https://www.fao.org/3/cb6410en/cb6410en.pdf

Food and Agriculture Organization of the United Nations (FAO) (2016) *Save and Grow in Practice—Maize, Rice, Wheat: A Guide to Sustainable Cereal Production'*. Available at: http://www.fao.org/3/i4009e/i4009e.pdf

Frenzel, P. (2000) 'Plant-associated methane oxidation in rice fields and wetlands', *Advanced Microbial Ecology*, vol. 6, pp. 85–114.

Gathorne-Hardy, A., Narasimha Reddy, D., Venkatanarayana, M. and Harris-White, B. (2016) 'System of rice intensification provides environmental and economic gains but at the expense of social sustainability: A multidisciplinary analysis in India', *Agricultural Systems*, vol. 143, pp. 159–168.

Hayashi, K., Liorca, L., Rustini, S., Setyanto, P. and Zaini, Z. (2018) 'Reducing vulnerability of rainfed agriculture through seasonal climate predictions: A case study on the rainfed rice production in Southeast Asia', *Agricultural Systems*, vol. 162, pp. 66–76.

Intergovernmental Panel on Climate Change (IPCC) (1996) *Climate change 1995: the science of climate change, Summary for Policy-makers'*. Contribution of working group 1 to the Second Assessment Report of the Intergovernmental Panel on Climate Change, Cambridge University Press, vol. 2, Cambridge, United Kingdom.

Intergovernmental Panel on Climate Change (IPCC) (2007) *Climate Change 2007: Impacts, Adaptation and Vulnerability'*. Fourth Assessment Report on Climate Change. Cambridge University Press, Cambridge, United Kingdom.

Intergovernmental Panel on Climate Change (IPCC) (2014) 'Climate Change 2014: Synthesis Report'. Contribution of Working Groups I, II and III to the Fifth Assessment Report of the Intergovernmental Panel on Climate Change, 88 pp.

Kern, J.S., Gong, Z.T., Zhang, G.L., Zhuo, H.Z. and Luo, G.B. (1997) 'Spatial analysis of methane emission from paddy soil in China and the potential for emission reduction', *Nutrient Cycling in Agroecosystem*, vol. 49, pp. 181–195.

Ketelaar, J.W., Morales-Abubakar, A.L., Van Du, P., Widyastama, C., Phasouysaingam, A., Binamira, J. and Dung, N.T. (2018) 'Save and Grow: Translating policy advice into field action for sustainable intensification of rice production', In: *Agricultural Development and Sustainable Intensification: Technology and Policy challenges in the face of Climate Change.* London: Earthscan, pp. 23–51.

Ketelaar, J.W., Abubakar, A.L., Phasouysaingam, A., Chanthavong,V., Dung, N.T., Flores-Rojas, M., Mishra, A. and Sprang, P. (2020) 'Save and Grow: Sustainable intensification of crop production and innovative market linkages for building resilient rural economies in rice landscapes in the Greater Mekong Sub-region', In: *The Bioeconomy Approach: Constraints and Opportunities for Sustainable Development*, Nagothu, US. (eds). ISBN 9780367335717. Routledge Taylor and Francis Group, United Kingdom.

Kirk, G.J.D. (2003) 'Rice root properties for internal aeration and efficient nutrient acquisition in submerged soil', *New Phytologist*, vol. 159, pp. 185–194.

MARD (2016) 'The 10 years journey of SRI in Vietnam'. Available at: https://vietnamsri.wordpress.com/2017/04/25/the-10-years-journey-of-sri-in-vietnam/

Mishra A. (2019a) '*Boosting Yields, Raising Incomes, and Offering Climate-Smart Options: The System of Rice Intensification Paves the Way for farmers to Become More Successful "Agripreneurs"*'. Available at: http://www.sri-lmb.ait.asia/downloads/SRI-LMB%20Final%20Report2019.pdf

Mishra, A. (2019b). 'Morphological and physiological root plasticity and its relationships with shoot growth of rice', In: *Root Biology – Growth, Physiology, and Functions*, Takuji O. (eds). doi: http://dx.doi.org/10.5772/intechopen.87099.

Mishra, A. and Kumar, P. (2009) 'Southeast Asia Regional Knowledge Exchange on SRI Producing More with Less Water', Asian Institute of Technology, Bangkok, Thailand, 109 pp.

Mishra, A. and Salokhe, V.M. (2010) 'The effects of planting pattern and water regime on root morphology, physiology and grain yield in rice', *Journal of Agronomy and Crop Science*, vol. 197, pp. 368–378.

Mishra, A. and Salokhe, V.M. (2011) 'Rice root growth and physiological responses to SRI water management and implications for crop productivity', *Paddy and Water Environment*, vol. 9, pp. 41–52.

Mishra, A. and Uphoff, N.T. (2013) 'Morphological and physiological responses of rice root and shoot to varying water regimes and soil microbial densities', *Archive of Agronomy and Soil Science*, vol. 59, pp. 705–731.

Mishra, A., Ketelaar, J., Uphoff, N. and Whitten, M. (2021) 'Food security and climate-smart agriculture in the lower Mekong basin of Southeast Asia: Evaluating impacts of system of rice intensification with special reference to rainfed agriculture' *International Journal of Agricultural Sustainability*, vol. 19, pp. 152–174. Available at: https://www.tandfonline.com/doi/full/10.1080/14735903.2020.1866852

Mishra, A., Kumar, P. and Noble, A. (2013) 'Assessing the potential of SRI management principles and the FFS approach in Northeast Thailand for sustainable rice intensification in the context of climate change', *International Journal of Agricultural Sustainability*, vol. 11, pp. 4–22.

Mishra, A., Whitten, M., Ketelaar, J.W. and Salokhe, V.M. (2006) 'The System of Rice Intensification (SRI): A challenge for science, and an opportunity for farmer empowerment towards sustainable agriculture', *International Journal of Agricultural Sustainability*, vol. 4(3), pp. 193–212.

Neue, H.U. (1993) 'Methane emission from rice fields', *Bioscience*, vol. 43, pp. 466–474.

Pretty, J., Attwood, S., Bawden, R., van den Berg, H., Bharucha, Z., Dixon, P., Flora, J., Gallagher, C.B., Genskow, K., Hartley, S.E., Ketelaar, J.W., Kiara, J., Kumar, V., Lu, Y., MacMillan, T., Marechal, A., Morales-Abubakar, A.L., Noble, A., Prasad, P.V. and Yang, P. (2020) 'Assessment of the growth in social groups for sustainable agriculture and land management', *Global Sustainability*, vol. 3, e23. Available at: https://doi.org/10.1017/sus.2020.19

Stoop, W.A., Uphoff, N.T. and Kassam, A. (2002) 'A review of agricultural research issues raised by the system of rice intensification (SRI) from Madagascar: opportunities for improving farming systems for resource-poor farmers', *Agricultural Systems*, vol. 71, pp. 249–274.

Stoop, W.A. (2011) 'The scientific case for system of rice intensification and its relevance for sustainable crop intensification', *International Journal of Agricultural Sustainability*, vol. 9(3), pp. 443–455.

Swanson, B.E. and Rajalathi, R. (2010). 'Strengthening agricultural extension and advisory systems: Procedures for assessing, transforming and evaluating extension systems'. *Agriculture and Rural Development Discussion Paper 45*. Washington, DC. The International Bank for Reconstruction and Development and World Bank.

Thakur, A.K. and Uphoff, N.T. (2017) 'How the system of rice intensification can contribute to climate-smart agriculture', *Agronomy Journal*, vol 109, pp. 1163–1183.

Thakur, A.K., Mandal, K.G., Mohanty, R.K. and Uphoff, N.T. (2021) 'How agroecological rice intensification can assist in reaching the Sustainable Development Goals'. Available at: https://doi.org/10.1080/14735903.2021.1925462

United Nations Environmental Programme (2021). 'Making peace with nature. A scientific blue print to tackle the climate, biodiversity and pollution emergencies'. Available at: https://www.unep.org/resources/making-peace-nature

United Nations Environmental Programme & Climate and Clean Air Coalition (2021) 'Global methane assessment: Benefits and costs of mitigating methane emissions'. Available at: Methane|Climate & Clean Air Coalition (ccacoalition.org) & Global Methane Assessment: Benefits and Costs of Mitigating Methane Emissions|UNEP – UN Environment Programme.

Wang, X.T. and Below, F.E. (1996) 'Cytokinins in enhanced growth and tillering of wheat induced by mixed nitrogen source', *Crop Science*, vol. 36, pp. 121–126.

Whitten, M.J. and Settle, W.H. (1998) 'The Role of the Small-scale Farmer in Preserving the Link between Biodiversity and Sustainable Agriculture', p. 187–207, In: Chou, C.H. and Shao, K.-T. (Eds.), *Frontiers in Biology: The Challenges of Biodiversity, Biotechnology and Sustainable Agriculture Proceedings of 26th AGM, International Union of Biological Sciences*, Taiwan 17–23 November 1997, Academia Sinica, Taipei 1998.

World Economic Forum (WEF) (2021) 'The Global Risks Report 2021', 16th Edition, *World Economic Forum*, ISBN: 978-2-940631-24-7, Available at: https://www3.weforum.org/docs/WEF_The_Global_Risks_Report_2021.pdf

6 Direct seeded rice

A potential climate-neutral and resilient farming system

Anjani Kumar, Amaresh Kumar Nayak, Mehreteab Tesfai, Rahul Tripathi, Sangita Mohanty, Shyamaranjan Das Mohapatra, Kiran Mohapatra and Udaya Sekhar Nagothu

Introduction

Rice (*Oryza sativa* L.) is one of the principal staple food crops, which ensures food and nutritional security to a large percent of the global population, especially in Asia. It is cultivated globally by more than 95 countries and occupies 11% of the world's arable land with an annual production of about 678 million tons (https://www.statista.com). Thus, rice is an important cereal crop on which global food security is dependent. The three major rice producing countries are China followed by India and Indonesia, which are also the three of the most populated countries. To satisfy the global food demand, food production needs to be increased by 70% by 2050 (Muthayya et al., 2014). The changing trend of farming and other land uses has decreased the availability of arable land for rice farming that limits further expansion of rice cultivation. Thus, increasing the rice productivity through intensification is one option to meet the growing demand for rice production. However, the intensification must be done in a sustainable manner with minimum environmental impacts (FAO, 2011, 2014, 2016). We need to learn from the past experiences, and any future strategy to increase rice production must be performed without further increasing greenhouse gas (GHG) emissions, particularly methane and nitrous oxide.

Conventionally, rice is grown by puddling the rice fields and transplanting the seedlings in the puddled land. The main advantage of this system includes increased nutrient mobility due to the continuous presence of standing water in the field, weed suppression and a stable yield. However, the anticipated climate change will negatively affect the precipitation pattern and crop water requirement, which in turn will affect the timely availability of irrigation water for sustaining rice production (IPCC, 2007). In Asia, anticipated water crisis is reported to be the root cause of 'physical water scarcity' for nearly 39 million ha of irrigated rice by 2025 (Tuong and Bouman, 2003). This would ultimately result in 30% decline in agricultural production by 2050 (Hossain and Siddique, 2015). Moreover, the presence of standing water in the traditional lowland paddy fields causes arsenic toxicity and emits significant amount of methane gas, which contributes to global

DOI: 10.4324/9781003273172-6

warming (Kumar et al., 2016). The other challenge is the migration of labour force from rural to urban areas, as a result of which the availability of labour for agricultural work is at stake in most parts of South and Southeast Asia.

Thus, rice cultivation is faced by two risks in Asia mainly from anticipated irrigation water scarcity due to climate change and variability and increased labour costs (Pandey and Velasco, 2005). In addition, there are other constraints including the availability of seeds and timely extension services support. Thus, rice demands continuous efforts to ensure a resource-efficient sustainable alternative system to cope with the vagaries of climate change and other growing risks. A viable alternative rice establishment technology, which can produce more grain with less labour and ensure optimal water use in an eco-friendly manner, would potentially be direct seeded rice (DSR).

The main objective of this chapter is to assess the performance of DSR compared to conventional puddled and transplanted rice, and further recommend strategies that can promote adoption and scaling up of DSR as a potential climate-neutral and resilient rice farming system (CNRFS). The chapter was arranged as follows: first, the introduction section that described the main challenges of rice production, followed by the principles and practices of DSR, its advantages and limitations. Then, the performance of DSR compared to conventional puddled and transplanted rice was assessed using evidence-based field data from India and other countries under both dry and wet conditions. The performance indicators that were evaluated included *grain yield, nutrient and water use, GHG emissions* and *socioeconomic benefits*. Finally, the chapter recommended strategies to promote adoption and scaling up DSR.

Direct seeded rice

Direct seeded rice (DSR) is an establishment technique in which seeds are sown directly in the main field rather than by transplanting seedlings from the nursery. This technique has several advantages; however, there are some challenges that limit the adoption of DSR (Table 6.1). The main challenges include *high weed infestation, root-knot nematode infestation, nutrient deficiency* (e.g., Bui et al., 2021) and relatively *lower grain yield* compared to conventional transplanted rice. Some of these problems can be effectively handled by following integrated approaches of nutrient, pest and weed management.

Methods of direct seeded rice

The main methods of DSR are dry-DSR, wet-DSR and water seeding (Figure 6.1).

i   *DSR with dry seeding (dry-DSR)*: The practice of dry seeding involves sowing of seeds in the field with optimum moisture conditions. Seeds are sown with pre-sowing irrigation to enhance a good seed germination rate and establishment before the onset of monsoons. This practice ensures timely crop establishment, which ensures higher productivity; in addition, it is less demanding in labour, water and energy compared to conventional paddy transplantation.

Table 6.1 Main advantages and disadvantages of DSR compared to transplanted puddled rice (TPR)

Advantages of DSR	References
• Saves irrigation water use by 30–50% and increases water productivity, if properly managed	Kumar et al. (2019) Field study (Figure 6.4)
• Reduces GHG emissions (mostly methane)	Singh et al. (2005)
• Saves labour requirement up to 60% (no transplanting, puddling and maintenance of standing water), thus reduces labour cost and renders higher net profit	Kumar and Ladha (2011) Field study (Figure 6.6)
• Maintains soil aggregates, reduces percolation losses, avoids formation of hard pans in the root zone and ensures favourable soil condition for succeeding crops	Sharma et al. (2003)
Disadvantages of DSR	
• Weed emergence in DSR puts a strong competition for nutrients, moisture, space and light, and results in reduction of economic yield	Bista and Dahal (2018)
• Nutrient uptake by rice roots under DSR is decreased due to change in nutrient dynamics, compared to TPR	Johnson et al. (2005)
• Higher occurrence of root-knot nematode which results in severe damage to rice in all ecologies	Prot et al. (1994)

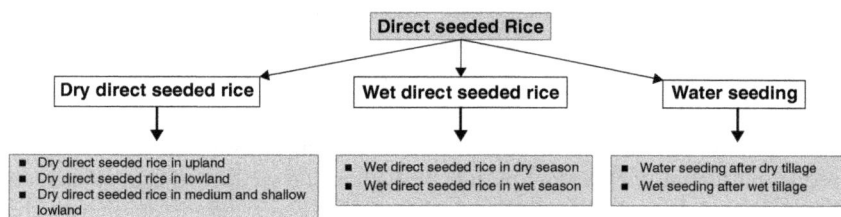

Figure 6.1 Different methods of DSR and its suitability in different rice ecologies.

ii *DSR with wet seeding (wet-DSR):* The main agronomic practices involved in wet seeding are the sowing of sprouted seeds (seed rate = 50 kg ha^{-1}) on the puddled bed with the help of drum seeder. The advantages of this method include reduction in labour costs and drudgery, besides, timely and better crop establishment (De Datta, 1986). The pre-requisites for successful wet-seeded rice are carefully levelled field and effective weed control (Balasubramanian and Hill, 2002).

iii *Water seeding:* Water seeding is mostly practised in irrigated lowlands where standing water of 5–15 cm is present. In this practice, land is dry ploughed, harrowed but puddling is avoided after dry tillage. Pre-germinated seeds are broadcast in the standing water of 10–15 cm depth; however, after wet tillage, puddling is practised.

In the following section, the performance of DSR was compared to conventional transplanted rice (TPR). The assessment was performed using some key indicators including *grain yield, water input and water use efficiency, nutrient use, GHG emissions, farm mechanization* and *benefit-cost ratio of DSR*. The results were discussed using the research findings from various studies conducted in different rice growing areas of India and other parts of Asia, under both dry and wet conditions. The data from the Government of Norway-funded Resilience project (www.resilienceindia. org) focusing on climate-smart rice production systems (2018–2022) were analysed to show the performance of DSR in Cuttack district, Odisha state of India.

Performance of DSR

i   *Grain yield*: Performance of DSR in terms of grain yield is dependent on many factors such as *climatic condition, crop establishment, precise management of inputs* (irrigation water and nutrients), *crop lodging* and *stakeholder's knowledge* (e.g., efficient use of machinery) *on* different farm operations (Rao et al., 2007).

Pilot demonstrations conducted under the Resilience project in Cuttack , Odisha (2018–2022) showed that grain yields of wet-DSR and dry-DSR were significantly higher than the TPR. The rice yield in DSR increased by 6. 71–13.3% in the wet season and by 6.45–11.5% in the dry season compared to TPR. This contrasts with a study conducted by Kumar and Ladha (2011) that showed reduction (9–28%) in grain yield under dry-DSR compared to conventional TPR. This may be due to different agro-ecological settings in which the two studies were carried out.

Grain yield under different crop establishment methods is summarized in Figure 6.2. In general, the grain yield under DSR varied from 3 t ha^{-1} (Farooq et al., 2009) to nearly 6 t ha^{-1} (Sharma et al., 2004). In most of the studies,

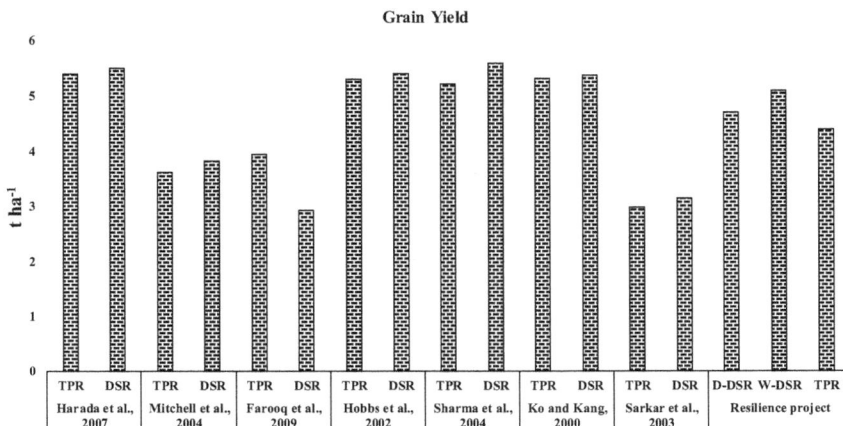

Figure 6.2 Rice grain yield under different crop establishment methods.

Table 6.2 Desirable traits of rice cultivars suitable for DSR

Desirable traits for DSR	References
• Lodging resistance	Mackill et al. (1996)
• Early seedling vigour for weed competitiveness	Zhao et al. (2006)
• Vigorous root system for better anchorage and soil moisture extraction	Pantuwan et al. (2002)
• Anaerobic germination	Ismail et al. (2009)
• Rapid shoot and root growth	Cui et al. (2002)
• Shorter duration of the crop	Dingkuhn et al. (1991)
• High crop growth rate during the reproductive phase	Kato et al. (2009)

the grain yield under DSR was higher than under TPR except in one study reported by Farooq et al. (2009). One of the reasons for the better performance of DSR was probably the desirable traits (Table 6.2) of the improved rice cultivars used and the good agronomic practices applied.

ii   *Water input and water use efficiency*: One of the critical factors for high water productivity in dry and wet seeded rice is precision water management. Maintaining aerobic conditions in the field is essential for promising crop stand establishment and high seedling vigour in early stages of dry seeding, whereas in wet seeded rice, precision water management is required for better performance of the applied herbicides and crop growth. Several studies were carried out during the last two decades, analysing the performance of water productivity and water savings under different rice systems. The research findings on total water inputs in rice under the different water management methods are summarized in Figure 6.3. The total water input ranged from about 3,500 mm under flooded and transplanted (Kato et al., 2009) to 500 mm under alternately submerged/non-submerged conditions (Belder et al., 2004) under different rice ecologies. The total water input under D-DSR, W-DSR in the Resilience project was lower than 1,000 mm except in TPR.

Experiments conducted under the Resilience India project reported that water saving was higher by 18–19.5% in wet-DSR and by 43–45% in dry-DSR over TPR (Figure 6.4). Similar findings were reported by Sharma et al. (2002) that observed 12–60% water savings under DSR and 13–30% under TPR. However, the water productivity under DSR in dry conditions (0.4–0.5 kg m^{-3}) was higher than that under DSR in wet conditions TPR (0.2–0.3 kg m^{-3}) (Figure 6.2). The water productivity can further be increased under DSR by precise land levelling using laser land leveller. This ensures uniform distribution of water, proper seed germination and weed control and good crop establishment, resulting in higher yield (7–24%) and irrigation water saving by 12–21% (Choudhary et al., 2002).

In DSR, precision irrigation practices like micro-irrigation, drip irrigation and other automated irrigation technologies can be used for enhancing water use efficiency. However, this requires additional investments that

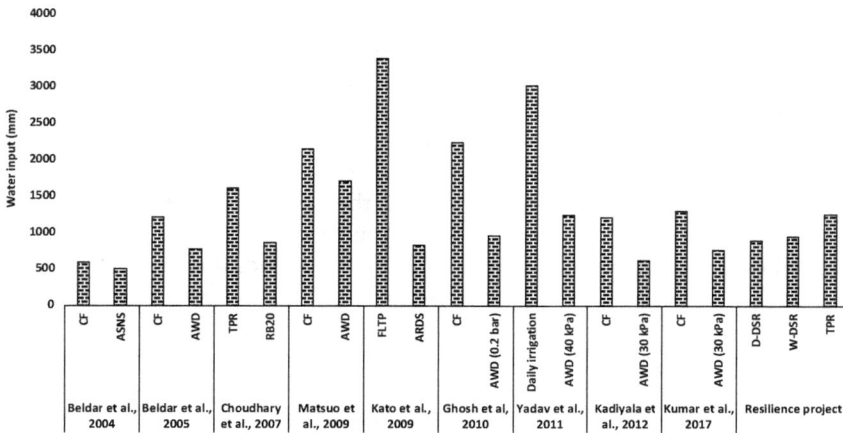

Figure 6.3 Representative studies reporting total water input in rice under different water regimes.

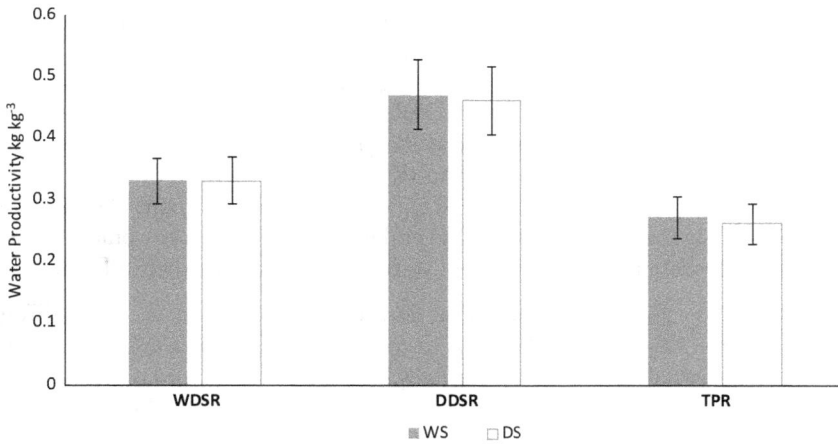

Figure 6.4 Comparison of cost of production and benefit-cost ratio under DSR and TPR.

smallholders cannot afford unless they are supported by subsidies from the government and maintenance of equipment. The Government of India has initiated country-wide programmes for upscaling precision irrigation systems for fruit crops, which may be eventually extended to rice (Agricoop, 2021). Drip and sprinkler irrigation technologies were found effective in saving irrigation water in rice up to 67%, and two-fold increase in the yield (Arns, 1999). Other advanced studies have shown that the application of artificial intelligence in automation of drip and sprinkler irrigation systems can further improve the application efficiency over surface irrigation method (Bhoi et al., 2021). However, drip and sprinkler methods are not commonly practised by farmers in the rice growing countries, due to difficulties in maintenance and initial establishment costs.

At the same time, we are observing a shift towards adoption of sensor-based technologies for site-specific and need-based application of irrigation water in India and other regions within the agriculture sector. Some of the sensor-based technologies used for scheduling irrigation include gypsum block sensor, time-domain reflectometry (TDR), frequency-domain reflectometry (FDR) and neutron probe sensors. Recently developed advancement in precision irrigation management is the development of Customized Colour-Coded Tensiometer (Kumar et al., 2021a) and the NRRI ARM sensor (Kumar et al., 2021b). Some of these sensors are being made farmer friendly and easy to handle and have the potential to save irrigation water, which can be up to 41% without any significant decline in the grain yield (Kumar et al., 2021a,b).

iii *Integrated nutrient and fertilizer management*: Site-specific or precision nutrient management is becoming important in rice and other cropping systems to improve productivity (Dobermann and Witt, 2004; Sapkota et al., 2016). This will not only reduce overuse of fertilizers but also reduce GHGs significantly. Proper dosage, and right time and method of application play an important role in nutrient management in rice. Otherwise, it can lead to losses of reactive nitrogen (N) through denitrification, volatilization and leaching, as observed under dry-DSR, which is higher compared to TPR (Davidson, 1991). As a result, the availability of plant nutrients such as nitrogen (N), phosphorus (P), potassium (K), sulphur (S), iron (Fe) and zinc (Zn) also reduced (Ponnamperuma, 1972), which hinders optimum plant growth and yield under DSR. There are several fertilizer management practices that can contribute to an improvement in the nutrient availability in DSR.

a *Split fertilizer application*: The dose of N fertilization used in DSR is higher than that used in TPR to compensate the higher losses of reactive N (Gathala et al., 2011). Normally, under DSR, one-third of the full dose of N, P and K is applied as basal dose, which enhances the fertilizer use efficiency by facilitating the availability of nutrients to the plants. The remaining two-thirds dosage of N is applied in equal splits at vegetative (active tillering) and reproductive (panicle initiation) stages (Kamboj et al., 2012). Such type of fertilizer application increases the grain yield and maximizes N use efficiency. More details about the dosage and timing of fertilizer application for DSR in different agro-ecological settings are provided in Box 6.1. Awareness about the fertilizer management suitable to DSR needs to be increased among farmers through regular trainings and information.

b *Green/brown manuring*: In conventionally tilled DSR, the use of chemical N fertilizer can be significantly reduced by applying green/brown manuring such as *Sesbania* (Farooq et al. 2021). The seeds of *Sesbania spp*

Box 6.1 Nutrient management recommendations for dry-DSR under different agro-ecological settings

a *Upland rice:* Well-decomposed farmyard manure or cow dung @ 2t acre^{-1} should be applied at the time of final land preparation, followed by a blanket dose of 24:12:12 kg of N:P:K acre^{-1}. For broadcast crop, the full dose of P and K during the final land preparation should be applied, and 75% of N after the first intercultural operation and rest 25% N at the panicle initiation (PI) stage. For line sown crops, 25% N + full P and K as basal dose and band placement is recommended and 50% N at three weeks after the first intercultural operation and rest 25% N at the PI stage. However, in sandy soils, K should be applied in two equal split doses (50% as basal and rest 50% at PI stage), whereas in acid soils, 50% P as SSP and rest 50% P through rock phosphate should be applied. Phosphorous solubilizing bacterial culture @ 25 g kg^{-1} is recommended to enhance availability of P in soil, whereas integration of 50% recommended dose fertilizer + *Gliricidia* @ 2.5 t ha^{-1} + Phosphate Solubilizing Bacteria @ 2.5 kg ha^{-1} + *Azotobacter* @ 2.5 kg ha^{-1} is recommended for autumn rice.

b *Deep water rice:* Blanket recommended dose is N:P:K @ 16–8–8 kg acre^{-1}. It is desirable to apply 8 kg P2O5 acre^{-1} during land preparation. The prevailing water regime does not allow application of nitrogenous fertilizer from late July to October. It is desirable to apply nitrogen fertilizer in one or two doses before water accumulates to depth of 5–10 cm in the field. Usually, fertilizer is placed in bands at sowing with seed-cum-fertilizer drill, hand plough or behind the country plough. For higher nitrogen use efficiency, 8 kg N acre^{-1} as basal and rest 8 kg is applied before flooding.

Source: Adapted from Saha et al. (2012).

@ 19.76 kg ha^{-1} are broadcast three days after rice sowing and allowed to grow for 25–30 days. It is then dried by spraying 2,4-D Ethyl Easter. In the case of broadcast rice, at the time of *beushening* (a traditional system of rice cultivation common in rainfed regions), harvested *Sesbania* foliage is incorporated in soil, whereas in the case of line sowing the incorporation is done at the time of manual weeding. This practice supplies about 14 kg of N per acre[1], adds organic matter to soil and helps in maintenance of overall soil health. Thus, a part of nitrogenous fertilizer (up to 25%) can be replaced by brown manuring. The occurrence of nematode infestation in DSR can be minimized by growing summer legume crops such as green gram in rice-wheat or green manuring of *Crotolaria juncea* L.

c *Fertilizer and seed treatment:* One of the important factors in sustainable rice production is integrated nutrient management, to improve the availability of nutrients applied through several fertilizer and seed treatment techniques. These techniques help not only in increasing the nutrient use efficiency but also reducing environmental pollution. The use of neem-coated urea (slow/controlled release of N fertilizers) and real-time N management significantly enhances the yield and nitrogen use efficiency. In the real-time N management practice, about 50% N application is adequately followed by split application based on the value shown by leaf colour chart, which is easy to use by farmers even without much literacy. In the case of K, split application (50% basal and 50% at panicle initiation stage) has proved to be more advantageous under the DSR condition (PhilRice, 2002). Seed treatment with suitable microbial inoculants helps DSR farmers for better nutrient cycling and enhancing the nitrogen use efficiency.

iv   *Greenhouse gas emissions:* Irrigated rice is one of the major sources of methane emissions (nearly 1.5% of total global GHG), a GHG which is more than 30 times as potent as carbon dioxide (Searchinger and Wiate, 2014). There are several studies that showed the practice of alternate wet and dry irrigation cycles in rice significantly helps in reducing CH_4 emission (Wang et al., 2012; Kumar et al., 2016). Some of the mitigation measures to GHG emissions from DSR based on previous research findings and the Resilience project results are presented in Figure 6.5. The findings showed that methane emission was lower in all DSR demo plots under wet/dry and zero tillage conditions than under the conventional TPR. The emissions were even lower under dry-DSR than under wet and/or zero tillage DSR. Careful land and water management in DSR with proper sequence of wetting and drying has the potential to reduce significant methane by 40–50%.

a   *Water management practices:* Water management practices like midseason drainage and intermittent drainage reduce emission of CH_4 more

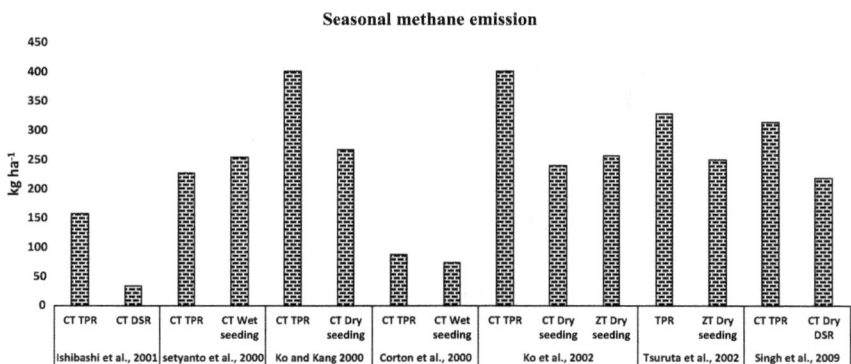

Figure 6.5 Research findings on methane emissions (kg ha^{-1}) from rice cultivation under TPR, DSR and ZT. CT = conventional tillage, TPR = transplanted puddled rice, DSR = direct seeded rice, ZT = zero tillage.

compared to flooded rice. However, earlier studies have also shown that midseason drainage and intermittent drainage increase N_2O emissions (e.g., Kumar et al., 2016). Li et al. (2011) reported that the timing and duration of midseason aeration affected the trade-off between CH_4 and N_2O emissions. It shows there is a potential for reducing GHG emissions from rice fields by adopting suitable water management practices (refer Chapter 3). Farmers need to be trained on the timings and method of water application to the rice fields for achieving net reductions in GHGs. Where farmers are dependent on canal irrigation, water management has to be coordinated by farmer groups and irrigation agencies collectively.

b *Crop establishment methods*: Under different crop establishment methods, Kumar and Ladha (2011) reported a higher N_2O-N emission under different dry-DSR practices (1.3–2.2 kg N_2O-N ha^{-1} in bed-dry-DSR and ZT-dry-DSR compared to 0.31–0.39 kg N_2O-N ha^{-1} in conventional TPR). These findings clearly indicate that although DSR has the capacity to reduce methane emission, it increases the N_2O emissions. The trade-off between CH_4 and N_2O emissions should be the basis for devising water management strategy under DSR for mitigating overall global warming potential (GWP) from rice fields. Crop establishment strategies that would reduce GWP will be ideal for fitting into the rice-based systems as alternatives for conventional practices.

Farmer-led demo trials during the years 2019–2021 (under the Resilience project) in the Tangi area of Cuttack district (Odisha state) reported that cumulative N_2O emission flux was between 1.1 and 1.5 kg N_2O-N ha^{-1} under DSR, which is higher than that under TPR by almost 42%. Similar results were found by Kumar and Ladha (2011) who reported higher N_2O emission flux (0.90–1.1 kg N_2O-N ha^{-1}) under conventional dry-DSR practice. The cumulative CH_4 emissions in WDSR and DDSR were 30.3% and 39% less than in TPR in wet season, whereas 36.8% and 47% less than in TPR in dry season (Table 6.2). The cumulative CO_2 emission flux in WDSR, DDSR and TPR were 1,323, 1,474 and 1,647 kg ha^{-1} in wet season and 1,520, 1,576, 1,656 kg ha^{-1} in dry season, respectively (Table 6.3). The cumulative GWPs were 20.5–21.6% less in WDSR and 22.7–23.3% less in DDSR compared to TPR. Other studies have reported similar results showing that the overall net effect of DSR in GWP was decreased by 16–33% compared to conventional rice production methods (Pathak and Aggrawal, 2012). Harada et al. (2007) reported that replacing conventional TPR by DSR in the Indo-Gangetic Plains can reduce the global warming potential by 25%. Thus, DSR provides an opportunity for the agriculture sector to contribute to mitigation efforts if included and supported within the Nationally Determined Contributions (NDCs) initiatives.

Farm mechanization

The small landholding size of many Indian farmers (<2 ha) is a major constraint to promote farm mechanization at the individual farm level. However, there is an

Table 6.3 Effect of greenhouse gas emissions, GWP on different crop establishment techniques

Treatment	CH_4 (kg ha^{-1})		CO_2 (kg ha^{-1})		N_2O (kg ha^{-1})		GWP (kg CO_2 ha^{-1})		% of GWP cut-off from FPR	
	WS	DS	WS	DS	WS	DS	WS	DS	WS	DS
WDSR	75.1	52.3	1,323	1,520	1.07	1.13	3,657	3,163	20.5	21.6
DDSR	65.5	43.1	1,474	1,576	1.48	1.46	3,552	3,099	22.7	23.3
TPR	108	82.8	1,647	1,656	0.92	1.10	4,598	4,038	–	–

WDSR – wet direct seeded rice, DDSR – dry direct seeded rice, TPR – transplanted puddled rice.
Source: Authors analysis of research data from Resilience project sites.

increasing trend of shifting towards farm mechanization due to labour shortage in India and other regions of South and Southeast Asia. The use of agricultural machinery like seed drills, power sprayers, combine harvesters and other agricultural implements is gradually increasing, which is an enabler for the successful adoption of DSR. Farmers adopting mechanized DSR can make use of power seed drills to reduce the seed rate significantly compared to the manual practice (Kumar and Ladha, 2011). Optimum farm mechanization has the potential to save time and labour, and curtail the cost of production and post-harvest losses. The downside of farm mechanization is increased use of fossil fuels that will contribute to GHG emissions. Hence, it is necessary to supplement with renewable energy source (e.g., solar) wherever possible, and hand-driven tillers and smaller machinery that are climate neutral.

Precision seed drills are gaining popularity due to their cost effectiveness and effective operation. The metering unit (roller type/spiral grooved) present in precision seed drills ensures the uniform seeding operation because they are easy to adjust for varying speed of operation. The seed rate of modern seed drill having fluted roller type metering mechanism varies from 25 to 30 kg ha^{-1} and for inclined plate and cup type metering mechanism, it varies from 15 to 20 kg ha^{-1}. Further lowering of seed rate can be achieved by using precision vacuum type metering unit. Seeding by precision seed drills resulted in an increase in seedling emergence by 7.8–8.7% (Zhang et al., 2017). Under wet-DSR condition, optimum seeding can be achieved by use of drum seeders. For better performance, the orifice diameter on its circumference, optimum speed of rotation, forward speed of operation and forward speed of operation to peripheral speed of drum needs must be standardized. Drought- or flood-tolerant seeds are not easily accessible and expensive for small farmers. Hence, lowering seed rate will also help farmers to reduce their input costs and help more farmers to access the climate-resilient seed material.

Despite the advantages of mechanization in DSR, the affordability and use of agricultural implements by smallholders continues to be low. This is due to the lack of availability of implements at an affordable price, inappropriate synchronization between demand and supply, lack of custom hiring centres, limited access to credit and lack of required technical knowledge and awareness on the efficient use of machinery for different farm operations. The Farmer Producer Organizations (FPOs) in India are setting up custom hiring centres (like case study districts of Assam and Cuttack in Odisha in the Resilience project) that will be useful for smallholders who do not have the capacity to purchase machinery on their own. Results from the Resilience project have shown that FPOs have the potential to support farmers to access farm machinery, seed material, training and marketing will help in upscaling of DSR.

DSR and advantages to women and youth

DSR provides several advantages to marginal and smallholder farmers in particular to women and youth. It saves labour requirement by up to 60% as it does not require transplanting, puddling and maintenance of standing water. Given the fact that women contribute a large share of the labour force for transplanting rice (and possibly the accompanying drudgery), adoption of DSR will give more time for women farmers to engage in other activities (e.g., household management, alternative income generating activities). Studies show that women are in favour of adopting DSR and are willing to learn and pay for technology such as DSR that reduces the agricultural workload (Khan et al., 2016). Youth can be trained and engaged in operating seed drills and combines increasingly used in DSR upscaling. Moreover, the practice of DSR enhances smallholder farmers' resilience and adaptation to climate change and variability, e.g., during times of drought and flooding. Thus, DSR has a potential to improve farmers' livelihood by increasing productivity and income as shown also in the studies conducted under the resilience project villages.

Economics of DSR

There are several determinants of cost of cultivation like labour wages, machine use, and irrigation cost (Tripathi et al., 2014). The profitability under DSR is location specific. Areas with cheaper labour cost and easy availability of irrigation water recorded higher benefit-cost ratio under DSR compared to other areas. There are several studies that showed economic prospects for DSR are highly promising (Pandey and Velasco, 2002). The cost of cultivation under DSR reduced by 45–48% compared to TPR due to crop establishment, followed by water management (Yaduraju et al., 2021). Other studies by Kumar and Ladha (2011) reported a reduction in the cost of cultivation by 6–32% under dry-DSR and by 2–16% under wet-DSR. This reduction in the production cost increases profitability mainly due to less tillage operations, cheap labour and irrigation water and more use of machines for different agricultural operations (Tripathi et al., 2014). Nevertheless, the cost for weed control is about 20–38% higher in DSR compared to TPR. Assuming other expenditures on crop cultivation in both TPR and DSR similar, adoption of DSR practice may result in total net saving of Indian rupees (INR) 9,114 to 10,192 per hectare. This implies that DSR adoption per million hectares of land would result in an economic benefit range of INR 10.0 billion (Yaduraju et al., 2021).

The inclusion of seed drill, power-operated boom sprayer and combine harvester in DSR (accessed by farmers through custom hiring centres) system contributed to reduced cost of cultivation by 25%. Moreover, the use of modern seed drills further decreased the seed rate under DSR by around 50% compared to TPR (Dhakal et al., 2019).

The pilot demos conducted at the Resilience project site in Odisha state, India, recorded a significantly higher benefit-cost ratio under DSR (1.6) compared to TPR (1.8). The use of seed drills in the DSR demos resulted in significant reduction in labour and fuel costs, which led to enhancing the farm profitability (see Figure 6.6).

Challenges and measures to upscale adoption of DSR

The practice of DSR is gaining popularity among rice farmers in water-scarce areas. Globally, about 23% of rice area is already under DSR (Rao et al., 2007; Kumar and Ladha, 2011). Adoption of DSR is increasing in rice growing areas of Southeast Asia (Singh et al., 2005; Bhandari et al., 2020). This is due to the advantages of DSR: it enhances water use efficiency, reduces methane emission, reduces labour cost, renders higher net profit and maintains soil physical properties (see Table 6.1).

However, farmers in India face challenges in the transformation and adoption of DSR. The main challenges are *shortage of climate resilient rice varieties* capable of performing on par with puddle transplanted conditions; *lack of effective herbicides, crop lodging, iron deficiency, nematode infestation, non-availability of appropriate machinery* for seeding rice; and *lack of awareness* on improved DSR production technology (Bhullar and Gill, 2020). Measures to overcome some of these challenges are discussed below:

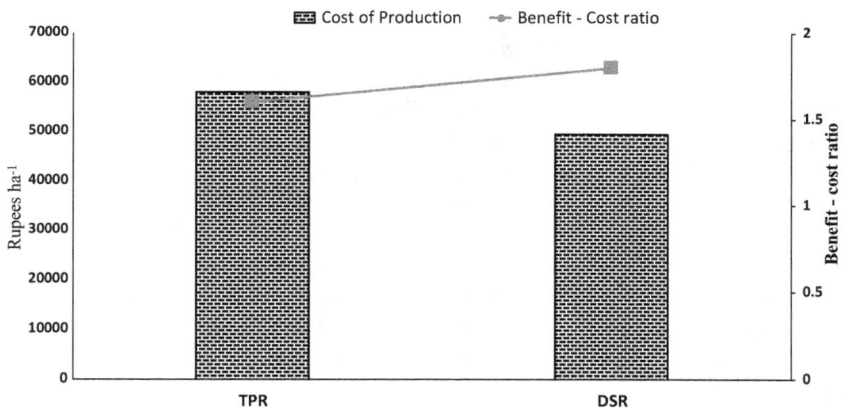

Figure 6.6 Comparison of cost of production and benefit-cost ratio under DSR and TPR. DSR = direct seeded rice, TPR = transplanted puddled rice.

i   *Shortage of climate resilient rice varieties*: In this regard, one success case story to mention is how the lack of good quality improved seeds was addressed by the Resilience project in Assam (India). Two climate resilient varieties were developed by the researchers and were introduced to farmers. As the demand for these seeds was increasing, there was a need to develop a public-private partnership between farmers and research community. The latter agreed to provide foundation seeds to farmers directly for on-farm multiplication under the supervision of the local research institute. This success has generated positive impacts among the farmers and increased the availability of improved seeds in the area. In addition, adaptive research should be carried out to produce rice varieties that adapt the local condition and tolerant to lodging problem.

ii *Lack of effective herbicides and nematode infestation*: There are many species of weed flora, of which hardy grassy weeds and sedges are more prevalent in DSR (Caton et al., 2003; Rao et al., 2007). The weed problems in DSR can be controlled using various measures that are context specific, as shown in Table 6.4.

   To implement the above-mentioned measures effectively, it requires training of farmers, increasing awareness and informing farmers about the right time and method for weed control during the cropping season and access to purchase the inputs.

iii *Nutrient uptake by rice roots and iron deficiency*: Several research efforts have been made on nutrient uptakes and iron deficiency in DSR (e.g., Zhang et al., 2019). However, there is limited knowledge to improve nutrient uptakes – in particular iron deficiency in DSR in specific ecologies. In general, measures such as applying site-specific nutrient management approach (refer Chapter 2) and growing rice cultivars with high iron contents could improve rice nutrient uptakes including iron in the soils.

iv *Non-availability of appropriate machinery* for seeding rice: This challenge could be addressed by introducing services hiring centres that provide access

Table 6.4 Examples of research findings related to effective weed control measures in DSR

Weed control measures	References
• Use of manual and/or self-propelled weeders can play a significant role in controlling weeds	Rao et al. (2007)
• Mulching with wheat residue @ 4 t ha^{-1} inhibits emergence of grassy weeds by 44–47% and broad leaf weeds by 56–72% in dry-DSR	Singh et al. (2007)
• Brown manuring reduces weed population at early stage due to its high growth rate and competition	
• Use of EMS mutant lines of Nagina 22 (N 22) was identified as Imazethapyr resistant. The trait has been transferred to Pusa Basmati 1121; non-genetically modified herbicide-tolerant rice varieties	Grover et al. (2020)

to farm machineries to smallholders, women and youth. For example, the Resilience project established a Custom Hiring Centre (CHC) in Assam (Golgahat district) to support farmers to provide farm machineries such as power tillers, paddy threshers, straw choppers, power weeders and winnowers at a subsidized hiring rate. The rent charge is used to cover the costs of maintenance and operation of CHC and the machineries. According to farmers, the timely availability of farm machinery has enabled them to perform the different farming activities efficiently.

v *Lack of awareness on DSR cultivation methods*: The virtual and/or physical Village Knowledge Centre (VKC) services (ICT-based digital tools) introduced in the Resilience project have increased farmers' awareness about DSR. Participatory extension services through VKCs and farmer-to-farmer learning have encouraged other farmers to practise DSR. The VKCs extension personnel working in collaboration with the state-run farmer training centres (KVKs) is helping in scaling up DSR.

Policy and institutional support for upscaling DSR

Extension activities can play a very important role in popularization of DSR, which includes training, demonstration of DSR in farmer's field, farmer-to-farmer learning, on-farm trials related to various potential problems faced by farmers and exposure visit of farmers to demo farms. In this regard, there are several opportunities at international, national and state levels that can enhance upscaling DSR practice in suitable rice ecologies.

Recent IPCC reports and other climate action initiatives globally are promoting DSR as one of the most promising climate-neutral farming systems (Sulaiman et al., 2018).

India is committed to fight the challenge of climate change considering CoP26 commitments. One of strategies/pathways prioritized in the policy agenda is to make Indian agriculture resilient and sustainable in a changing climate. DSR is having the potential for reducing GHG emission and is being included in the national missions that can help in achieving the target of net-zero emissions by 2070, which India has pledged as per the CoP26 at Glasgow, Scotland (Padhee and Whitbread, 2022). DSR has been prescribed as efficient in terms of input use like water, labour and energy when compared to traditionally perceived water-guzzling crop like paddy. In this regard, one of the salient recommendations that emerged during the First Indian Rice Congress held in December, 2020 in India recognized the potential of DSR as one of the promising climate-neutral and sustainable rice production option (Nayak et al., 2020).

Some states in India such as Punjab are targeting to bring 1 million ha rice cultivation under DSR technique. Efforts are underway in upscaling CSA technologies that include DSR with the support of KVKs and other stakeholders outside the project areas.

Conclusions

This chapter analysed the performance of DSR on grain yield, nutrient management, water use, GHG emissions and socioeconomic benefits and compared the results with TPR under different rice ecologies. DSR can be a viable alternative rice system to address the future challenges in the drier and warmer climate scenarios, provided enabling policy and institutional support are in place.

In general:

- DSR is promising as it overcomes the problem of labour and water scarcity.
- Adoption of DSR will benefit not only commercial farmers but also smallholders
- DSR reduces agricultural workload and provides women and youth the opportunity to move into more remunerative on-farm and off-farm employment.
- Further, application of modern biotechnology and development of effective crop, water and soil management practices has the potential to optimize crop yields under DSR.
- DSR can curtail methane emissions by 44%, and the cumulative GWP by 25% compared to TPR if properly managed.

A coordinated effort is needed to promote adoption and upscaling of DSR. In this regard, researchers must focus on developing rice varieties that have early vigour, robust root architecture and weed competitiveness during early stages of crop growth. Extension workers must increase demonstrations and training farmers and other stakeholders about the practices of DSR and its benefits. Policy makers and government agencies should support the uptake and promotion of DSR practice in the rice farming systems through investments and marketing support measures. Finally, farmers practising DSR should be compensated for their contribution to reduce GHG emissions.

Acknowledgements

The chapter has benefited from the results of the Resilience project field trials in the Odisha state of India. The authors would like to thank field enumerators and lead farmers involved in the field demos. At the same time, they acknowledge the support of the Norwegian Ministry of Foreign Affairs/The Norwegian Embassy, New Delhi for funding and support to the Resilience project in India (2018–2023).

References

Agricoop (2021) 'Annual report, 2020–2021'. Available at: https://agricoop.nic.in/sites/default/files/Web%20copy%20of%20AR%20%28Eng%29_7.pdf
Arns, W. and Arns H (1999) 'Growing rice with pivots–a step towards water conservation, Rio Srande do sue region, South Porazil', Available at: http://www.irri-ar.com.ar/

Balasubramanian, V. and Hill, J.E. (2002) 'Direct seeding of rice in Asia: Emerging issues and strategic research needs for the 21st century'. In: Pandey, S., Mortimer, M., Wade, L., Tuong, T.P., Lopez, K. and Hardy, B. (Eds.) *Direct Seeding: Research Strategies and Opportunities*, pp. 15–39, International Rice Research Institute, Los Banos, Philippines.

Belder, P., Bouman, B.A.M., Cabangon, R., Lu, G., Quilang, E.J.P.L.Y., Spiertz, J.H.J. and Tuong, T.P. (2004) 'Effect of water-saving irrigation on rice yield and water use in typical lowland condition', *Asian Agricultural Water Management*, vol. 65, pp. 193–210.

Belder, P., Bouman, B.A.M., Spiertz, J.H.J., Peng, S., Castaneda, A.S. and Visperas, R.M. (2005) 'Crop performance, nitrogen and water use in flooded and aerobic rice', *Plant Soil*, vol. 273, pp. 167–182.

Bhullar, M.S. and Gill, J.S. (2020) 'Farmer's apprehensions about DSR and possible solutions', *The Times of India*, 13 July 2020. p. 3.

Bhoi, A., Nayak, R.P., Bhoi, S.K., Sethi, S., Panda, S.K., Sahoo, K.S. and Nayyar, A. (2021) 'IoT-IIRS: Internet of Things based intelligent-irrigation recommendation system using machine learning approach for efficient water usage', *Peer Journal of Computer Sciences*, vol. 7: e578

Bhandari, S., Khanal, S. and Dhakal, S (2020) 'Adoption of direct seeded rice over puddled-transplanted rice for resource conservation and increasing wheat yield', *Reviews in Food and Agriculture*, vol. 1(2), pp. 59–66.

Bista, B. and Dahal, S. (2018) 'Cementing the Organic Farming by Green Manures', *International Journal of Applied Sciences*, vol. 6, pp. 87–96.

Bui, H.X., Oliveira, C.J., Desaeger, J.A. and Schroeder, N.E. (2021) 'Rice root-knot nematode *Meloidogyne graminicola* (Nematoda: Chromadorea: Tylenchida: Meloidogynidae: Meloidogyne). Entomology and Nematology Department'. Available at: https://edis.ifas.ufl.edu/publication/IN1350

Caton, B.P., Cope, A.E., and Mortimer, M. (2003) 'Growth traits of diverse rice cultivars under severe competition: implications for screening for competitiveness', *Field Crop Research*, vol. 83, pp. 157–172.

Choudhary, M.A., Gill, M.A., Kahlown, M.A. and Hobbs, P.R. (2002) 'Evaluation of resource conservation technologies in rice-wheat system of Pakistan: Developing an action program for farm-level impact in rice-wheat systems of the Indo-Gangetic plains, p. 112.

Cui, K., Peng, S., Xing, Y., Xu, C., Yu, S. and Zhang, Q. (2002) 'Molecular dissection of seedling-vigor and associated physiological traits in rice', *Theoretical and Applied Genetics*, vol. 105, pp. 745–753.

Davidson, E.A. (1991) 'Fluxes of nitrous oxide and nitric oxide from terrestrial ecosystems', *The American Society of Microbiology*, Washington, DC.

De Datta, S.K. (1986) 'Technology development and the spread of direct-seeded flooded rice in Southeast Asia', *Experimental Agriculture*, vol. 22, pp. 417–426.

Dhakal, R., Bhandari, S., Joshi, B., Aryal, A., Kattel, R.R. and Dhakal, S.C. (2019) 'Cost-benefit analysis and resource use efficiency of rice production system in different agriculture landscapes in Chitwan district', Nepal. *Archives of Agriculture and Environmental Science*, vol. 4, pp. 442–448.

Dingkuhn, M., Penning, de Vries, F.W.T., De Datta, S.K. and van Laar, H.H. (1991) 'Concepts for a new plant type for direct seeded flooded tropical rice. Direct-seeded flooded rice in the tropics', Selected Papers from the International Rice Research Conference, Seoul, Korea, 27–31 August 1990, pp. 17–38.

Dobermann, A. and Witt, C. (2004) 'The evolution of site-specific nutrient management in irrigated rice systems of Asia', In *Increasing productivity of intensive rice systems through site-specific nutrient management* (eds. Dobermann, A., Witt, C. and Dawe, D.) 410 (Science Publisher Inc. and International Rice Research Institute, IRRI).

Farooq, M., Basra, S.M.A., Ahmad, N. and Murtaza, G. (2009) 'Enhancing the performance of transplanted coarse rice by seed priming', *Paddy Water Environment*, vol. 7, pp. 55–63.

Farooq, M., Ullah, N., Nadeem, F., Nawaz, A. and Siddique, K.H. M. (2021) 'Sesbania brown manuring improves soil health, productivity, and profitability of post-rice bread wheat and chickpea', *Experimental Agriculture*, vol. 57, pp. 145–162.

FAO (Food and Agriculture Organization) (2011) 'Save and grow. A policymaker's guide to the sustainable intensification of smallholder crop production', Rome.

FAO (Food and Agriculture Organization) (2014) 'Building a common vision for sustainable food and agriculture: Principles and approaches', Rome.

FAO (Food and Agriculture Organization) (2016) 'Save and Grow in practice maize rice wheat. A Guide to sustainable cereal production', Rome.

Gathala, M.K., Ladha, J.K., Kumar, V., et al. (2011) 'Tillage and crop establishment affects sustainability of South Asian rice-wheat system', *Agronomy Journal*, vol. 103, pp.961–971.

Grover, N., Kumar, A., Yadav, A.K., Krishnan, S.G., Ellur, R.K., Bhowmick, P.K., Vinod, K.K., Bollinedi, H., Nagarajan, M., Viswanathan, C. and Sevanthi, A.M.V. (2020) 'Marker assisted development and characterization of herbicide tolerant Near Isogenic Lines of a mega Basmati rice variety, Pusa Basmati 1121', *Rice*, vol.13 (1), pp.1–13.

Harada, H., Kobayashi, H. and Shindo, H. (2007) 'Reduction in greenhouse gas emissions by no-tilling rice cultivation in Hachirogata polder, northern Japan, Life-cycle inventory analysis', *Soil Science and Plant Nutrients*, vol. 53, pp. 668–677.

Hossain, M.A. and Siddique, M.N.A. (2015) 'Water-A limiting resource for sustainable agriculture in Bangladesh', *EC Agriculture*, vol. 1(2), pp. 124–137.

IPCC (2007) 'Climate change 2007: the physical science basis. Contribution of working group I', In: Solomon, S., Qin, D., Manning, M., Chen, Z., Marquis, M., Averyt, K.B., Tignor, M. and Miller, H.L. (Eds.), *Fourth Assessment Report of the Intergovernmental Panel on Climate Change*. Cambridge University Press, Cambridge, United Kingdom.

Ismail, A.M., Ella, E.S., Vergara, G.V. and Mackill, D.J. (2009) 'Mechanisms associated with tolerance to flooding during germination and early seedling growth in rice (*Oryza sativa* L.).' *Annals of Botany*, vol. 103, pp. 197–209.

Johnson, S.E., Lauren, J.G., Welch, R.M. and Duxbury, J.M. (2005) A comparison of the effects of micronutrient seed priming and soil fertilization on the mineral nutrition of chickpea (*Cicer arietinum*), lentil (*Lens culinaris*), rice (*Oryza sativa*) and wheat (*Triticum aestivum*) in Nepal', *Experimental Agriculture*, vol. 41, pp. 427–448.

Kamboj, B.R., Kumar, A., Bishnoi, D.K., Singla, K., Kumar, V., Jat, M.L., Chaudhary, N., Jat, H.S., Gosain, D.K., Khippal, A., Garg, R., Lathwal, O.P., Goyal, S.P., Goyal, N.K., Yadav, A., Malik, D.S., Mishra, A. and Bhatia, R. (2012) 'Direct seeded rice technology in Western Indo-Gangetic Plains of India: CSISA experiences', CSISA, IRRI and CIMMYT, p. 16.

Kato, Y., Okami, M. and Katsura, K. (2009) 'Yield potential and water use efficiency of aerobic rice (*Oryza sativa* L.) in Japan', *Field Crop Research*, vol. 113, pp. 328–334.

Khan, T., Kishore, A. and Joshi, P.K. (2016) 'Gender dimensions on farmers' preferences for direct-seeded rice with drum seeder in India', IFPRI Discussion Paper 01550.

Kumar, A., Nayak, A.K., Mohanty, S. and Das, B.S. (2016) 'Greenhouse gas emission from direct seeded paddy fields under different soil water potentials in Eastern India', *Agricultural Ecosystems and Environment*, vol. 228, pp. 111–123.

Kumar, A., Nayak, A.K., Das, B.S., Panigrahi, N., Dasgupta, P., Mohanty, S., Kumar, U., Panneerselvam, P. and Pathak, H. (2019) 'Effects of water deficit stress on agronomic and physiological responses of rice and greenhouse gas emission from rice soil under elevated atmospheric CO_2', *Science of the Total Environment*, vol. 650, pp. 2032–2050.

Kumar, A., Nayak, A.K., Tripathi, R., Mohanty, S. and Nayak, P.K. (2021a) 'Customized color coded tensiometer for scheduling irrigation in rice', NRRI technology bulletin-154, Published by Director, ICAR – NRRI, Cuttack.

Kumar, A., Nayak, A.K., Hanjagi, P.S., Tripathi, R., Mohanty, S. and Panneerselvam, P. (2021b) 'NRRI-ARM sensor—A tool for real time soil moisture monitoring., NRRI technology bulletin-175. Published by ICAR – NRRI, Cuttack, India.

Kumar, V. and Ladha, J.K. (2011) 'Direct seeding of rice: recent developments and future research needs', *Advances in Agronomy*, vol. 111, pp. 297–413.

Li, D., Liu, M., Cheng, Y., Wang, D., Qin, J., Jiao, J., Li, H. and Hua, F. (2011) 'Methane emissions from double-rice cropping system under conventional and no-tillage in Southeast China', *Soil Tillage Research*, vol. 113, pp. 77–81.

Mackill, D.J., Coffman, W.R. and Garrity, D.P. (1996) 'Rainfed lowland rice improvement', *International Rice Research Institute*, Los Banos, Philippines.

Muthayya, S., Sugimoto, J.D., Montgomery, S. and Maberly, G.F. (2014) 'An overview of global rice production, supply, trade, and consumption', *Annals of the New York Academy of Sciences*, vol. 1324 (1), pp. 7–14.

Nayak, A.K., Baig, M.J., Samantaray, S., Bhattacharyya, P., Mohapatra, S.D., Mondal, B., Kumar, A., Adak, T., Raghu, S., Molla, K.A., Parameswaran, C., Hanjagi, P.S.H., Gowda B.G., Pradhan, A.K., Sinha, S.K. and Sethi, S.K. (2020) 'Extended summaries, rice research and development for achieving sustainable development goals, 1st Indian Rice Congress – 2020, Association of Rice Research Workers, ICAR-National Rice Research Institute, Cuttack – 753 006 (Odisha), India.

Padhee, A.K. and Whitbread, A (2022) 'Indian agriculture: The route post-CoP 26', Available at: https://www.downtoearth.org.in/blog/climate%20change/amp/indian-agriculture-the-route-post-cop-26-81154

Pandey, S. and Velasco, L. (2002) 'Economics of direct seeding in Asia: Patterns of adoption and research priorities', In: Pandey, S., Mortimer, M., Wade, L., Tuong, T.P., Lopez, K., Hardy, B. (Eds.), *Direct Seeding: Research Strategies and Opportunities*, International Rice Research Institute: Los Banos, Philippines, pp. 3–14.

Pandey, S. and Velasco, L. (2005) 'Trends in crop establishment methods in Asia and research issues', Rice is Life: Scientific Perspectives for the 21st Century. Proceedings of the World Rice Research Conference, 4–7 November 2004, Tsukuba, Japan, pp. 178–181

Pantuwan, G., Fukai S., Cooper M., Rajatasereekul, S. and O'Toole, J.C. (2002) 'Yield response of rice (*Oryza sativa L.*) genotypes to different types of droughts under rainfed lowlands. Plant factors contributing to drought resistance', *Field Crops Research*, vol. 73, pp. 181–200.

Pathak, H. and Aggarwal, P.K. (2012) 'Low carbon technologies for agriculture: a study on rice and wheat systems in the Indo-Gangetic Plains', Indian Agricultural Research Institute, New Delhi, pp. 12–40.

PhilRice (The Philippine Rice Research Institute). (2002) 'Integrated nutrient management for rice production', In *Rice Technology Bulletin No. 45*, p. 24. The Philippine Rice Research Institute (PhilRice): Maligaya, Nueva Ecija, Philippines.

Ponnamperuma, F.N. (1972) 'The chemistry of submerged soils', *Advances in Agronomy*, vol. 24, pp. 29–96.

Prot, J.C., Villanueva, L.M. and Gergon, E.B (1994) 'The potential of increased nitrogen supply to mitigate growth and yield reductions of upland rice cultivar UPL Ri-5 caused by *Meloidogyne graminicola*'. *Fundamentals of Applied Nematology*, vol. 17, pp. 445–454.

Rao, A.N., Johnson, D.E., Sivaprasad, B., Ladha, J.K. and Mortimer, A.M. (2007) 'Weed management in direct-seeded rice', *Advances in Agronomy*, vol. 93, pp. 153–255.

Saha, S., Rao, K.S., Jena, S.R., Pande, K., Das, L., Behera, K.S. and Panda, B.B. (2012) Agro-techniques of wet direct-sown summer rice. DST Project Technology Bulletin No.1. Published by Director, Central Rice Research Institute, Cuttack – 753006.

Sapkota, T.B., Majumdar, K., Khurana, R., Jat, R.K., Stirling, C.M. and Jat, M.L. (2016) 'Precision nutrient management under conservation agriculture-based cereal systems in South Asia' In: Nagothu, U.S (Ed.), *Climate Change and Agricultural Development: Improving Resilience through Climate Smart Agriculture, Agroecology and Conservation*. New York: Routledge, pp: 131–160, ISBN: 978-1-138-92227-3.

Searchinger, T. and Waite, R. (2014) 'More rice, less methane', Available at: https://www.wri.org/insights/more-rice-less-methane.

Sharma, P.K., Singh, R.P. and Kumar, R. (2004) 'Effects of tillage on soil physical properties and crop performance under rice-wheat system', *Journal of Indian Society of Soil Sciences*, vol. 52, pp. 12–16.

Sharma, P.K., Ladha, J.K. and Bhushan, L. (2003) *Soil Physical Effects of Puddling in Rice-wheat Cropping Systems. Improving the Productivity and Sustainability of Rice-wheat Systems: Issues and Impacts*, ASA Special Publication, vol. 65, 2003.

Sharma, P.K., Bhushan, L., Ladha, J.K., *et al.* (2002) 'Crop-water relations in rice-wheat cropping under different tillage systems and water-management practices in a marginally sodic medium textured soil', *Agronomy Journal*, vol. 64, pp. 521–524.

Singh, Y., Singh, G., Johnson, D. and Mortimer, M. (2005) 'Changing from transplanted rice to direct seeding in the rice-wheat cropping system in India Rice is Life': *Scientific Perspectives for the 21st Century, Proceedings of the World Rice Research Conference, 4–7 November 2004*, pp. 198–201. Tsukuba, Japan.

Singh, S., Ladha, J.K., Gupta, R.K., Bhushan, L., Rao, A.N., Sivaprasad, B. and Singh, P.P. (2007) 'Evaluation of mulching, intercropping with *Sesbania* and herbicide use for weed management in dry-seeded rice (*Oryza sativa* L.)', *Crop Protection*, vol. 26, pp. 518–524.

Sulaiman, R.V., Chuluunbaatar, D. and Vishnu, S. 2018. *Upscaling Climate Smart Agriculture. Lessons for Extension and Advisory Services*. FAO ROME.

Tripathi, R.S., Raju, R. and Thimmappa, K. (2014) 'Economics of Direct Seeded and Transplanted Methods of Rice Production in Haryana', *ORYZA International Journal of Rice*, vol. 51, 70.

Tuong, T.P. and Bouman, B.A. (2003) 'Rice production in water-scarce environments: Water productivity in agriculture': *Limits and Opportunities for Improvement*, vol. 1, pp. 13–42.

Wang, J., Zhang, X., Xiong, Z., Khalil, M.A.K., Zhao, X., Xie, Y. and Xing, G. (2012) 'Methane emissions from a rice agroecosystem in South China: effects of water regime, straw incorporation, and nitrogen fertilizer', *Nutrient Cycling in Agroecosystems*, vol. 93 (1), pp. 103–112.

Yaduraju, N.N., Rao, A.N., Bhullar, M.S., Gill, J.S. and Malik, R.K. (2021) 'Direct-Seeded Rice (DSR) in India: New opportunities for Rice Production and Weed management in post-COVID-19 pandemic period', *Weeds Journal of the Asian-Pacific Weed Science Society*, vol. 3(2), pp. 30–48.

Zhang, X., Liu, H., Zhang, S., Wang, J. and Changzhou Wei, C. (2019) 'NH$_4$–N alleviates iron deficiency in rice seedlings under calcareous conditions'. Available at: www.nature.com/scientificreports/

Zhang, M., Xiwen, L., Wang, Z., Wang, B. and Zhenlin, X. (2017) 'Optimization design and experiment of profiling and slide board mechanism of precision rice hill-drop drilling machine', *Transactions of the Chinese Society of Agricultural Engineering*, vol. 336, pp.18–26.

Zhao, D.L., Atlin, G.N., Bastiaans, L. and Spiertz, J.H.J. (2006) 'Cultivar weed competitiveness in aerobic rice: Heritability, correlated traits, and the potential for indirect selection in weed-free environment', *Crop Science*, vol. 46, pp. 372–380.

7 Carbon-neutral farming solutions in rice farming systems in Europe

Stefano Monaco, Patrizia Borsotto, Roberto Cagliero, Chiara Bertora, Omedé Gabriele, Maite Martínez-Eixarch and Laura Bardi

Introduction

In Europe, rice cultivation in terms of area and production has been stable between 1994 and 2020. The amount of rice produced in 2019 was about 2.8 million tonnes from a total cultivated area of about 428,000 ha, which represented 0.27% of world's rice production and 0.41% of world's harvested area (FAOSTAT, 2019). Due to temperature and water needs, rice in Europe is only cultivated in southern regions and in specific areas or districts, mainly in Italy (53.2% of total area as of 2020) and Spain (23.9%), and to a lesser extent in Greece (8.4%), Portugal (6.2%), France (3.6%), Bulgaria (2.9%), Romania (1.4%) and Hungary (0.7%). The production provides nearly 60% of internal rice consumption needs in Europe; the remainder 40% (approximately 1.2 million tonnes of milled rice per year, mainly Indica/long grain rice) is imported, especially from Pakistan, Thailand, Myanmar and Cambodia (EC, 2019a).

Though the rice production in the European Union (EU) is comparatively smaller to the total global production, the rice farming systems in some of the European regions have a long tradition and have important economic, cultural and landscape relevance at local and regional scale. The current research on improving sustainability and carbon neutrality of rice farming in Europe can therefore be useful to other rice growing regions in the world. This is one of the main reasons for including this chapter in the book.

In the EU, rice is normally cultivated under flooding conditions, sown in spring and harvested in autumn. Average yields range between 4 and 8 tonnes per hectare, also depending on cultivars that mostly belong to Japonica rice varieties. Flooded paddy landscapes providing habitats for many organisms including migratory birds are important for biodiversity conservation and artificial wetland maintenance in Europe. This agroecosystem is similar to other rice growing regions, where several environmental problems can be observed due to high use of fertilizers and agrochemicals, leading not only to pollution of soil and water but also to greenhouse gas (GHG) emissions (Kraehmer et al., 2017). Moreover, climate change (CC) further increases the vulnerability of these specialized farming systems due to the rise in water shortage, new pests and diseases and soil salinity in coastal regions.

DOI: 10.4324/9781003273172-7

GHG emissions from agriculture and rice farming systems in the EU

Considering all sources and sectors, except for LULUCF,[1] the total GHG emissions decreased by about one-third in EU since 1990. A total of 3.6 Gt of CO_2 eq emissions was estimated in 2019 (EEA, 2019), which is about 10% of the global GHG emissions (UNFCCC, 2019). According to the EEA, this is mainly due to the implementation of the EU and national policies and measures that have contributed to the decrease of GHG emissions in almost all sectors, particularly in energy supply, industry and the residential sector, while emissions from agriculture have increased in recent years. The EU has set new targets of 55% reduction by 2030 compared with 1990 level and of achieving a climate-neutral economy by 2050, which will need substantial efforts across all the sectors of the economy. The agricultural sector has contributed to 12.7% of the total GHG emissions in 2019, mostly methane (CH_4) and nitrous oxide (N_2O) (EEA, 2019). The shares of CH_4 and NO_2 emitted from the agricultural sector are highly relevant in the EU, because they correspond to 53.7% and 74.6% of total CH_4 and NO_2 emissions, respectively. Although rice in the EU is usually cultivated under flooded conditions, which causes high amount of CH_4 emissions during crop growing season at field scale, the contribution of CH_4 from paddy fields to the total CH_4 emissions is only 1.2%, due to the very limited area of rice cultivation, while the main sources are represented by ruminants' enteric fermentation (80.7%) and manure management (17.4%). Nevertheless, as the contribution of CH_4 emissions from rice paddies to global warming is relevant at global scale, the effort for its reduction concerns the entire global community, as recognized by the EU in its strategy to reduce methane emissions (EC, 2020). Concerning N_2O emissions, the direct and indirect N_2O emissions from agricultural soils are relevant, representing 88.6% of the sector in 2019, while the remaining share originates from manure management (11.2%). Policies on climate impacts or rice assume that N_2O emissions from this crop are negligible or very small, in fact, less than 10% of the total emissions, and none of the rice growing countries include them in their national inventories. Nevertheless, water-saving techniques, which have been developed and spread to reduce water use and CH_4 emissions from paddies, may increase N_2O emissions through aerobic-anaerobic cycling, which favours nitrification and less complete denitrification. A more detailed account of the aerobic-anaerobic cycling and its impact on N_2O emissions is given in Chapter 3.

Besides the information on the contribution of the agriculture and rice sector to climate change through IPCC-supported national GHG inventory methodology, several other methods have been proposed for calculating the magnitude of impact per kilogram of food products and for understanding where the impacts are concentrated within the production chain from field to supermarket. Life cycle assessment (LCA) is the most utilized method for evaluating the environmental impacts of processes and products through indicators such as global warming potential (GWP) and carbon footprint. In an LCA study carried out in the Vercelli district, which is one of the most important rice producing area in Northern Italy, Blengini and Busto (2009) estimated that in the baseline rice farming system, direct emissions from the field were the first source of GWP with 68% of contribution, followed by fertilizer production (9%) and product transportation (6%). They also investigated the effects of alternative rice farming systems such as organic farming and water-saving techniques that provided interesting results. For organic farming, a lower impact per hectare but a higher GWP per unit mass of product was assessed, which increased by 20%, as expected due to lower yields. As organic farming has other important beneficial effects, such as biodiversity conservation and environmental pollution reduction, which are also major concerns in large rice producing regions of Asia, the research strategies for maintaining high yield and low GWP per product in the organic rice system have become more important in recent years. The use of alternatives such as water-saving techniques, which could also cause yield reductions and even concerns about its feasibility and negative trade-offs at large scale, has nonetheless the potential to decrease GWP by about 50%.

EU policy aspects and economics drivers

Climate action is one of the EU's key priorities and at the heart of European Green Deal initiative (EC, 2019b). Climate and energy are one of the five fields of action to which the EU has assigned specific objectives to be achieved within the framework of the Europe 2030 strategy, the aim of which is to fill the gaps in the EU growth model, thus creating the conditions for smart, sustainable and inclusive growth. The EU pursues these objectives through a combination of programmes and financial support measures, including the European Structural and Investment Funds, contributing to the thematic objectives of supporting the transition to a climate-neutral economy in all sectors and promoting climate change adaptation, risk prevention and management.

Concerning the agricultural sector, a stronger focus on green issues is one of the main emphases through innovations of the future Common Agricultural Policy (CAP), and it was already implemented partially in the 2014–2020 programming period (Matthew, 2012). Under the first pillar, direct payments were subject to cross-compliance, i.e. the observance of constraints aimed at environmental protection, food safety, animal welfare and the maintenance of land in good condition, in addition to commitments to the so-called "green payment" or greening. Under the second pillar there was specific support for those farmers who, voluntarily exceeding the baseline of cross-compliance and greening, decided to adopt more sustainable production practices on the farm. Besides CAP, the organization of the Rural Development Programmes (RDP) foresees that resources are concentrated on several measures in order to achieve synergistically certain common strategic priorities. It was possible to highlight Priority 5 that aims at "Promoting resource efficiency and the shift to a low-carbon and climate resilient economy in the agri-food and forestry sector". This broader policy priority was broken down into five specific areas of intervention, which are known as Focus Areas (FAs).

In the 2014–2022 program period, RDPs cover a specific Focus Area (FA 5E – fostering carbon conservation and sequestration in agriculture and forestry) designed to support actions that achieve carbon conservation and sequestration in agriculture and forestry, although other interventions dedicated to the same objective may have been introduced in other FAs (Figure 7.1). However, a recent report from the European Court of Auditors found that CAP funds attributed during 2014–2020 to climate action had little impact on agricultural emissions and that most mitigation measures had low potential to mitigate climate change while CAP rarely has been financing measures with high climate mitigation potential (ECA, 2021). The EU has the target to mobilize 4 billion euros of public expenditure, i.e. 2.5% of the total amount of the EU budget, to support 4 million hectares of agricultural and forest land contributing to carbon sequestration and conservation, which accounts for 2% of the EU agricultural and forest land. The commitment made by the different member states is variable and the contribution of each single RDP to meet the European target can vary greatly: in some cases, the RDPs have not even activated Focus Area 5E, in others the implementation of the Focus Area was quite ambitious. It shows that though policies have been developed in the EU to address climate mitigation, their implementation has not met the expectations.

Mitigation potential targets and methodologies to achieve them

To achieve 2030 and 2050 GHG emission targets, the EU should implement effective and verifiable measures to reduce net GHG emissions, including carbon dioxide (CO_2) removal actions. The agricultural sector could have a key role in this strategy with several options for mitigation, including the recognition

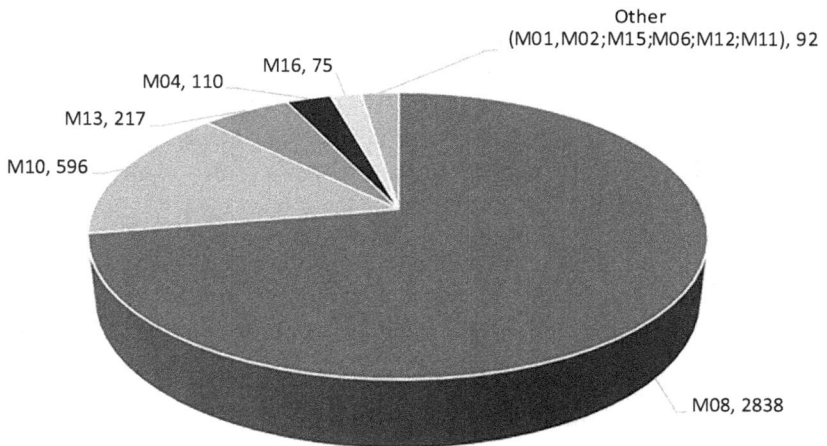

Figure 7.1 Planned public expenditure on Focus Area 5E per measure (millions of euro). Source: ENRD 2016.

and implementation of nature-based solutions as options for achieving climate neutrality by 2050. Moreover, sequestration methodologies could also have strong positive trade-offs, such as resilience increase of land and farming systems to climate change and other risks, improvement of resource use efficiency and enhancement of soil health and biodiversity. Mitigation potentials of the EU farming systems, together with feasibility, impact and cost effectiveness of monitoring, reporting and verification (MRV) system of possible measures and options, have been addressed in several research studies and in specific policy reports (Martineau et al., 2016; Leip et al., 2017; COWI, 2021). The concept of "Carbon farming" has been gaining more relevance in the EU context and its implementation is expected to start in the coming years (EC, 2021). It refers to the management of carbon pools, flows and GHG fluxes at farm level, with the aim of reducing net GHG emissions for climate change mitigation. This objective can be achieved through measures that involve the management of land, crops, soil and livestock.

Examples of mitigation actions at farm level to manage carbon and GHG fluxes, identified as relevant within the EU context, are reported in Table 7.1 with estimates of their potential impact. Martínez-Eixarch et al. (2021) assessed 22 different actions reported in the table and their mitigation potential and found that 11 could have substantial impact, with more than 5,000 kt of CO_2 eq per year of reduction at EU level. Among them, eight actions were related to carbon sequestration due to land use, land use change or crop production; two were related to mitigation of N_2O emissions from fertilizer application; and one (carbon audits) was a means of identifying relevant actions at farm business level. While the potential contribution to climate mitigation should be the first aspect to consider in any assessment of future potential schemes, other factors need to be considered as well. The permanence of the effect of the climate action on carbon pool and GHG fluxes, and the associated risk of reversal processes such as land management change or a catastrophic event such as fire, must be adequately considered, together with the additionality of the effect that without the action taken the results would not have occurred. Moreover, it is necessary to evaluate the risk of the so-called "carbon leakage", which is the displacement of the emission to another location, by emitting activities increase or land use changes caused by the considered mitigation action. Another important aspect is related to the evaluation of the achievement of the expected results in terms of carbon stock increase or GHG emission reduction, and in terms of the associated uncertainty, reliability and costs of the applied methodology that could be based on direct measurements, assessment tools and/or modelling. These are challenges that should be considered while planning for mitigation measures and their actual implementing and monitoring at each farm level in the coming years. This could be even a bigger challenge in scattered and small-size farms in Asia or Africa, where farmers' climate literacy could be limited. Any lessons learnt from the EU small farms in this context could be relevant for small-scale farming systems elsewhere.

Table 7.1 Potential GHG savings from various measures in practice to reduce EU agriculture and land use emissions

Group	Mitigation action	Range of values utilized for the assessment	Mitigation potential (kt CO_2 eq/y)	
			Min	Max
Land use	Conversion of arable land to grassland	2.2–7.3 t/ha/y CO_2e sequestered in soil	2,670	8,850
	New agroforestry potential	0.15–0.88 t/ha/y CO_2e sequestered in soil	257	1,560
	Wetland/peatland conservation/restoration	1.3–8.2 t CO_2e /ha/y	1.61	10.1
	Woodland planting	1.47–1.83 t/ha/y CO_2e sequestered in soil	2,570	3,210
	Preventing deforestation and removal of farmland trees	0.73–7.3 t/ha/y CO_2e sequestered in soil	1,079	10,790
	Woodland, hedgerows, woody buffer strips and trees on agricultural land (management)	0.37 t/ha/y CO_2e sequestered in soil	5,500	5,500
Crop production	Reduced tillage	0.0059–0.0180 t CO_2e /ha/y for fuel saved	104.5	324
	Zero tillage	0.0121–0.0359 t CO_2e /ha/y for fuel saved	809	2,467
	Leaving crop residues on the soil surface	0.11–2.2 t/ha/y CO_2e	133.3	2,670
	Ceasing to burn crop residues and vegetation	0–0.512 Mt per member state CO_2e per year	880	
	Use cover/catch crops	0.88–1.47 t/ha/y CO_2e sequestered in soil	10,460	18,100
Nutrient and soil management	Soil and nutrient management plans	0.033–0.159 t CO_2e /ha/y from N_2O reduction	2,130	21,300
	Use of nitrification inhibitors	0.003–0.017 t CO_2e /ha/y from N_2O reduction	29,700	89,100
	Improved nitrogen efficiency	0.033–0.159 t CO_2e /ha/y from N_2O reduction	2,130	21,300
	Biological N fixation in rotations and in grass mixes	0.006–0.042 t CO_2e /ha/y from N_2O reduction	6,390	12,400

Livestock production	Livestock disease management	1,060	10,600
	Use of sexed semen for breeding dairy replacements	481	1,730
	Breeding lower methane emissions in ruminants	21.3	44.4
	Feed additives for ruminant diets	860	1,730
	Optimized feeding strategies for livestock	280	850
Energy	Carbon auditing tools	12,900	19,900
	Improved on-farm energy efficiency	3,090	6,280

Adapted from Martineau et al., (2016).

Analysis of climate actions at farm scale

In Italy, rice production is mainly concentrated in the north-western Po Valley, covering 220,000 ha (Ente Nazionale Risi, 2020). In Vercelli, Pavia and Novara provinces, rice cultivation constitutes about 81% of the land use and has a long tradition of cultivation dating back to the 15th century. Currently, it represents one of the most intensive rice production systems in the world. In this area, rice is permanently flooded during most part of the growing season, from April-May to August-September. It is mostly cultivated in specialized farms, with yields ranging from 6 to 7 t ha^{-1} (Ente Nazionale Risi, 2020), depending on the type of cultivar (i.e. round, Long A or Long B grain size), high level of mechanization (i.e. 628 kW/farm) (RICA, 2019) and high use of chemical inputs, especially herbicides but also nitrogen fertilizers, fungicides and insecticides. This type of farming system has caused several negative environmental effects, such as water pollution, biodiversity loss and soil fertility depletion, which, together with high costs of production for inputs, the fluctuation of the global price of commodities and the reduction of protection measures of EU agricultural policies, has strongly questioned its economic and environmental sustainability in recent years. Climate change scenarios have further been enhancing the vulnerability of the rice farming system in the Po Valley, due to the changes in the rainfall and snow regimes and the increased frequency of extreme drought and water shortage events.

In this context, it is necessary to modify agricultural practices, and adopt new technologies and nature-based innovations at farm and field scale to increase rice farming system sustainability and resilience to climate change. Moreover, this sector can play an important role in mitigation actions by reducing GHG emissions and increasing C sequestration in land. Organic agriculture could represent a potential solution in favour of both environmental and economic sustainability. However, this farming system is a novelty and a challenge for the Italian rice sector, meaning significant changes in the agricultural practices (e.g. introduction of crop rotation, new practices for weed management) and uncertainty in terms of productive performance (Orlando et al., 2020). This could also mean a risk to farmers due to low productivity in organic rice production, despite its market demand in the EU.

For this reason, a research project (Risobiosystems 2017–2020) funded by the Italian Ministry of Agriculture, Food and Forest Policies (MIPAAF) and carried out by CREA, together with other Italian research institutions and universities, aimed at analysing national organic rice farming systems through a multidisciplinary approach, considering key agronomic, environmental and socio-economic aspects (SINAB, 2017). One of the research activities focused on studying a group of organic rice farms in Northern Italy to identify the main strategies, in terms of weed control, crop rotation and other agronomic and innovative aspects, supported by evaluating their productive, environmental and economic performances. The sample of farms was constituted by a network created by the University of Milan for the implementation of participatory research activities (Orlando et al., 2020), which was expanded during the project.

The sample farms were in the Provinces of Vercelli, Pavia and Novara, and specialized in rice production. Most of them adopted the organic rice farming system on the entire farm area, while the rest had a mixed management: a part of the farm was intended for conventional production and the rest under organic farming. The mixed organic-conventional farming is allowed by the EU rules, and it is common in the rice organic systems of Northern Italy. This is due to management problems, especially related to the technical difficulties of managing large farm areas under organic farming (e.g. additional time required due to mechanical interventions), and economic reasons (e.g. risk of income reduction due to uncertain and lower yields of organic rice and alternative crops). In the farm sample, the conversion from conventional to organic farming was rather recent: two of the farms shifted to organic in 2015, and two farms were under organic production since 2000. With regard to land use and crop rotation, rice cultivation represented about one-third of the area under organic farming, but the crop rotations adopted were generally done with few crops that often were functional to rice cultivation (i.e. cover crops for mulching or green manure). The choice of crops to be grown in rotation alternating with rice is one of the most critical issues for the organic rice system in Northern Italy, due to soil type and water management of paddy fields that could limit the growth of alternative crops, the specialization of farms and the economic value of the rice compared to other crops.

In a sub-sample of 12 farms, a study on the impact of organic practices and other innovative solutions on GHG emissions, carbon sequestration and carbon footprint was carried out (Table 7.2). Within this sample group, in each mixed-type farm, the organic and conventional lands were analysed separately. Organic rice areas were managed differently with the introduction of different innovative agricultural practices (Table 7.3).

Main strategies were especially targeted to weed control, which is the major challenge in the Italian rice production system. In organic agriculture this is even worse due to the prohibition of herbicides use: weed incidence was reported as the

Table 7.2 Rice farms analysed for carbon footprint evaluation

Farming type	ID	Utilized agricultural area	Geographic region	Soil characteristics
		ha		
Mixed	Org01	95	Vercellese plain (VC)	Silty loam, sub-acid or acidic
	Conv01	145		Silty loam, sub-acid
Conventional	Conv02	210	Vercellese plain (VC)	Silty loam, sub-acid or acidic
Mixed	Org03	44	Novarese plain (NO)	Different soil textures, alfisol
	Conv03	116		
Mixed	Org04	75	Novarese plain (NO)	Different soil textures, alfisol
	Conv04	100		
Mixed	Org05	13	Novarese plain (NO)	Clay soil, difficult water drainage
	Conv05	37		
Organic	Org06	40	Baraggia vercellese (VC)	Heavy soil, difficult water drainage, acid
Organic	Org07	46	Low plain in Lomellina (PV)	Sandy loam, sub-acid
Organic	Org08	122	Baraggia vercellese (VC)	Heavy soil, difficult water drainage, acid
Organic	Org09	21	Baraggia vercellese (VC)	Heavy soil, difficult water drainage, acid
Organic	Org10	21	Baraggia vercellese (VC)	Heavy soil, difficult water drainage, acid
Organic	Org11	30	Baraggia vercellese (VC)	Heavy soil, difficult water drainage, acid
Organic	Org12	98	Low plain in Lomellina (PV)	Sandy loam

Source: Based on own data collection through farmer interviews and field survey.

main cause of yield variability and yield gap of organic paddies. The use of green mulching from different cover crops was widespread in a major part of farms of the sample, aimed to limit weed development through the physical obstruction obtained by cover crop residue biomass and, to some extent, the phytotoxic action of chemical compounds (e.g. organic acids) produced during its fermentation in flooding water. In this technique, the rice seeds are broadcast or directly seeded on the cover crop (e.g. ryegrass and/or vetch), left standing or rolled or chopped just before field flooding, inducing biomass fermentation processes. This technique was carried out solo in the entire rice farm area or in combination with a different strategy, combining false seedbed preparation and mechanical weed control. Green mulching is in general associated with dry seeding in farmland with light soils and limited water availability, and it is carried out with a weeder harrowing used several times at pre-seeding, during pre-emergence and after the three leaves stage. In area with heavy soil with problems of water drainage, false seedbed and

Table 7.3 List of innovative agricultural practices applied in the subgroup of rice farms

Farm	Innovative rice cultivation practices	Rice area ha	Paddy yield t ha^{-1}
org01	Mulching with cover crop biomass, false seedbed, green manure, manual weed control	60	4.3
org03	Mulching with cover crop biomass, mulching with biofilm	44	3.7
org04	Mulching with cover crop biomass, minimum tillage	39	3.2
org05	Mulching with cover crop biomass	4	3.8
org06	Mulching with cover crop biomass	28	3.4
org07	Mulching with cover crop biomass	40	3.7
org08	Mulching with cover crop biomass, agroforestry	60	4.1
org09	False seedbed	21	3.6
org10	Mulching with cover crop biomass	21	3.6
org11	False seedbed	8	5.2
org12	AWD, transplant, interrow cultivator	30	5.0
conv01	Low chemical input use	145	4.4
conv02	Green manure	210	6.5
conv03	Winter flooding, mulching with cover crop, straw removed	116	7.5
conv04	High chemical input use, precision farming, minimum tillage	100	9.3
conv05	High chemical input use, precision farming	36.6	7.9

Source: Based on own data collected from field work.

mechanical weed management was carried out under saturated soil conditions (i.e. "puddling technique"). An improvement of mechanical weed control was carried out by seeding rice rows at 30 cm and using an interrow cultivator with satellite control as observed in the sample farm org12. Other innovative practices tested in the sample of farms included alternate wetting and drying (AWD) technique (Monaco et al., 2021), winter flooding, minimum tillage, mulching with biofilm, transplanting and agroforestry. Low chemical inputs for herbicides and fertilizers (conv01) and green manure (conv02) were tried out on the areas under conventional farming. The precision agriculture technique carried out in conv04 and conv05 rice area consisted in variable fertilization rate, based on prescription map, with the aim of yield maximization.

The carbon footprint of farms was calculated using the Cool Farm Tool© (CFT) with the data collected through interviews. The tool which was developed by the Cool Farm Alliance is an online tool used for the calculation of various agro-environmental indicators (CFA, 2019). Among the different tools available for calculating the "carbon footprint" on a farm (Whittaker et al., 2013), the CFT was selected for the following reasons: (i) it includes all emission sources; (ii) it allows us to manage the estimation for several farms simultaneously; and, above all, (iii) it presents a specific modality for rice cultivation (Hillier et al., 2011). The soil carbon sequestration was assessed using RothC model, one of the most popular models for this research topic (Coleman et al., 1996). Due to the anaerobic conditions of paddy soil caused by flooding, which slows down organic matter decomposition, it was necessary to apply proper coefficients of the decomposition/and mineralization rate to obtain reliable results for paddy, using the version developed by Shirato et al. (2005) in RothC-26.3. The application of CFT and RothC focused on rice cultivation and the outputs were reported using both surface and product as reference unit. The results of carbon footprint and carbon sequestration assessment are reported in Figure 7.2, while the detail of all other sources of GHG emissions except methane from paddy is reported in Figure 7.3.

Total carbon footprint and field emissions

The results of the assessment showed that methane emissions from paddy fields represented the highest GHG emission source in all cases considered in the study, with an average value of 83% of total emissions, that is 5.2 out of 6.2 t CO_2 eq ha^{-1}, and 1.2 out of 1.3 t CO_2 eq per t of paddy produced. The different strategies for weed control reached on average a similar total GHG emissions level when referred to the surface unit: 5.6 and 5.7 t CO_2 eq ha^{-1} for green mulching (org05-06-07-10) and false seedbed (org09-11), respectively, while a different level was assessed when referred to the product unit, that is 1.6 and 1.3 t CO_2 eq per t of paddy for green mulching and false seedbed, respectively. This difference is mainly due to lower yields with green mulching. The Org12 farm showed lower emissions (i.e. 4.5 t CO_2 eq ha^{-1}) due to the application of AWD associated with false seedbed technique, which confirmed that the duration of soil anaerobic conditions is one of the most important mitigation options. This was also highlighted by farms conv02, conv04 and conv05, which showed a lower amount of field CH_4 emissions (i.e. 3.9 t CO_2 eq ha^{-1}) due to a lower number of days of flooding with respect to organic management, for drainage periods necessary to spread herbicides and fungicides. In addition, the adoption of winter flooding practice in conv03 determined the highest GHG emissions assessed at surface level (i.e. 12.5 t CO_2 eq ha^{-1}), partially compensated by very high yield when referred to product unit. The productive performance greatly influenced the total GHG emissions based on product unit, leading to lower values, equal to 0.89, 0.75, 0.58 and 0.89 t CO_2 eq per t in org12, conv02, conv04 and conv05, respectively. The latter farms achieved the highest yield levels by applying variable rate fertilizers based on yield map.

The other sources of GHG emissions (Figure 7.3) were related to the use of fertilizers, which caused direct field emissions, the management of residues, the industrial production of chemical input, especially fertilizers, and the use of energy both for field operations and for grain drying. The production and use of mineral fertilizers, especially nitrogen fertilizers (that require large amounts of energy to be produced and induce high direct N_2O emissions from the soil), represented a relevant emission source under conventional farming (0.7 t CO_2 eq ha^{-1}), as well as for org01 and org12 (0.4 t CO_2 eq ha^{-1}) due to organic fertilizers. Among the organic farming cultivation options, the application of false seedbed

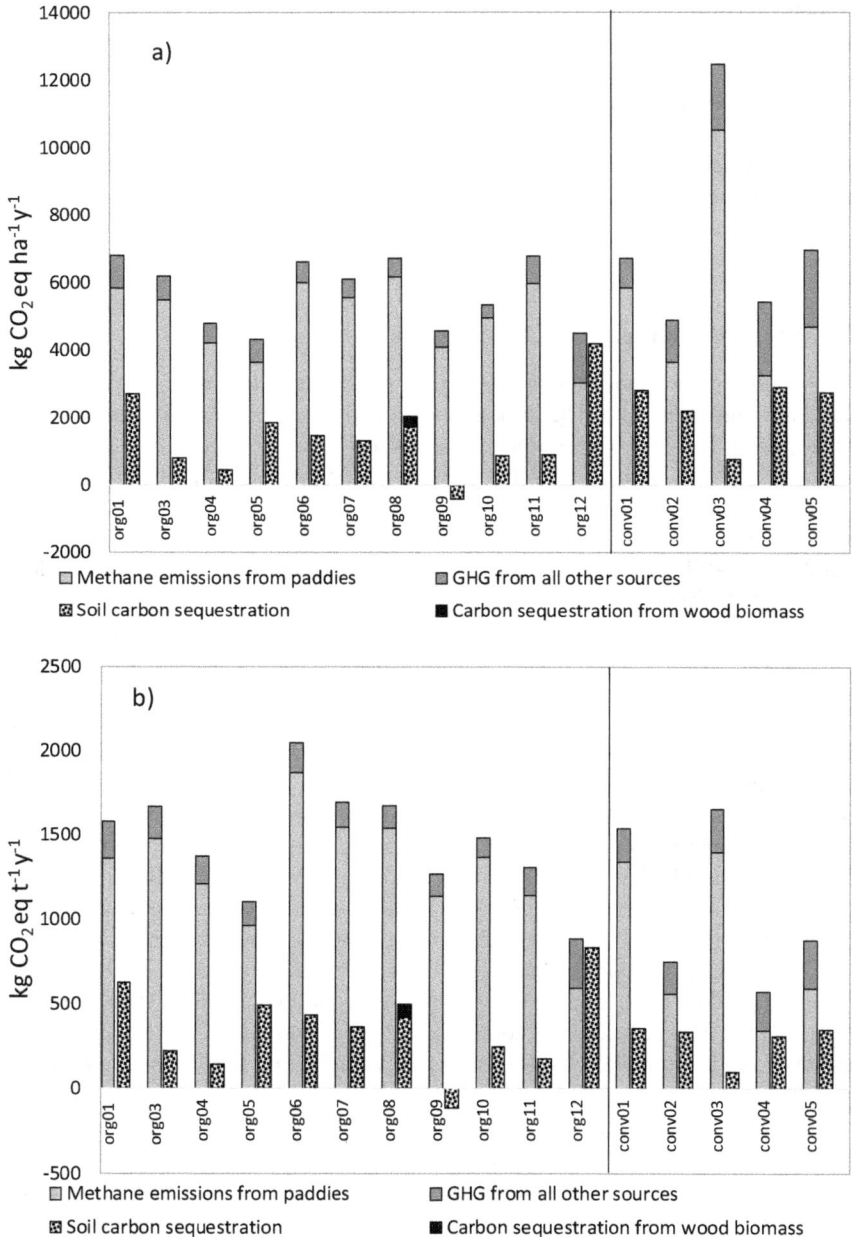

Figure 7.2 GHG emissions and soil carbon sequestration under different rice management referred to the surface (a) and rice product (b) unit.

Source: Authors' own data calculated with Cool Farm Tool and RothC using farms survey data.

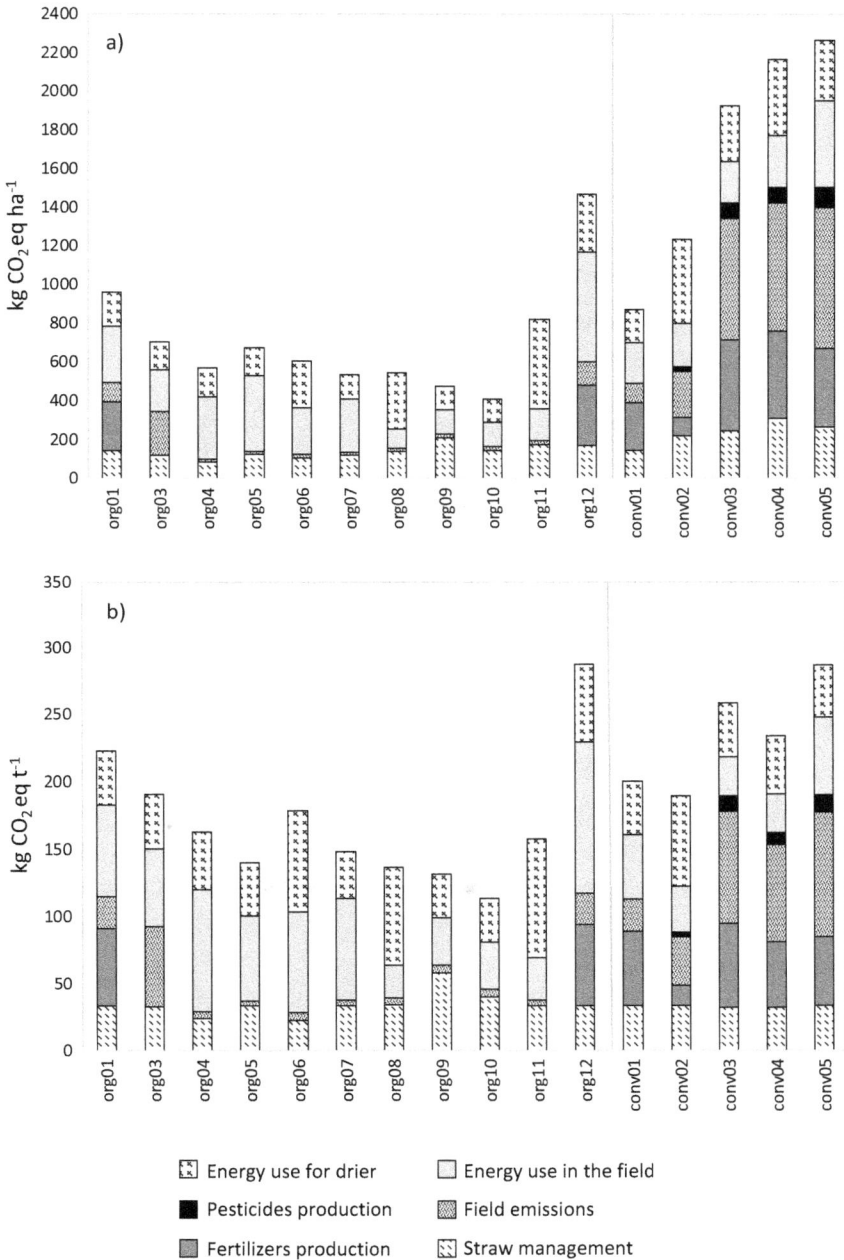

Figure 7.3 GHG emissions from all other sources except methane emissions from paddies under different rice managements referred to the surface (a) and rice product (b) unit.

and mechanical weed control with high energy consumption represented a relevant emission source in the sample farm org12 (1.5 t CO_2 eq ha^{-1}).

Carbon sequestration

Considering soil carbon sequestration calculated by RothC-26.3 model, which is reported in Figure 7.2, the current management techniques would lead to an increase of 0.2% in organic carbon in 20 years and will sequester 1.7 t CO_2 ha^{-1} per year on average for all farms, which corresponds to 27.6% of total GHG emissions. Conventional management generally showed higher values than organic management. This was probably due to the calculation by the model of higher carbon input from straw produced and returned to the soil. In the sample farm conv03, where straw was removed, the soil carbon sequestration was lower. Moreover, there was a large variability of results among organic farming, probably due to the different soil characteristics and the initial level of soil organic matter. Nevertheless, the sample farm org09 was the only management option that showed a clear trend of soil organic content depletion, which could be attributed to low yield (and straw return) and absence of cover crops. Low values of carbon sequestration were assessed when vetch was used as cover crop compared to ryegrass, due to low biomass production and C input. In the sample farm org12, the result obtained was very high, but this is mainly due to the very low initial carbon content. Positive variations largely depended on the initial conditions of the soil and therefore comparisons with other situations can be difficult. The information per se is useful for comparing different practices in a given area. Carbon sequestration due to tree planting under agroforestry was observed on one of the sample farms (org08) and was calculated using the CFT web application. Agroforestry is a practice that has a high potential to store carbon effectively, yet its contribution in this case was observed to be limited. The rather low result may be due to several factors, including low tree density and inadequacy of CFT on carbon input estimates from temperate climate tree crops.

Discussion and relevance to other rice growing regions

Rice crop in Europe is mostly cultivated under permanent flooding conditions, which induces anaerobiosis in soil and consequently CH_4 emissions, which represents the highest contribution to carbon footprint of rice production. Then, the mitigation options for reducing CH_4 emissions mainly deal with reducing the flooding duration under both conventional and organic farming.

In European rice farming systems, the effects of different water management techniques on GHG emissions and the resulting global warming potential (GWP) have been studied in Italy and Spain. Peyron et al. (2016) compared both dry seeding and intermittent irrigation (dry seeding followed by intermittent irrigation throughout the growing season) with permanently flooded irrigation system. While intermittent irrigation showed a larger mitigation capacity (>70% reduction of GWP) than dry seeding (56%), with almost negligible CH_4

emissions throughout the growing season, it induced significant yield losses; on the contrary, dry seeding, which is a common practice in Italy, introduced and widespread in the last years, maintains or even increases yield (Miniotti et al., 2016). Similarly, Bertora et al. (2018) found a significant reduction of 69% of CH_4 emissions in rice fields with dry seeding in comparison to permanently flooded fields. However, it is worth to note that the CH_4 mitigation capacity of dry seeding when compared to sowing in water may depend on the reference cultivation system. In Spain, Martínez-Eixarch et al. (2018) found very low emissions of CH_4 during early vegetation stages under permanently flooded rice cultivation, likely because of the large CH_4 emissions during the preceding winter that depleted labile organic matter. Therefore, an insignificant mitigation effect of dry seeding under cultivation systems could be expected wherein fields are flooded over winter. On the contrary, the AWD technique tested in Spain (Martínez-Eixarch et al., 2021) significantly reduced CH_4 emissions up to 95%: so, it could effectively reduce the overall GWP of the rice fields, even when the increase in N_2O emissions is accounted for.

The main organic substrate providing methanogenesis is straw. In a pot experiment, Bertora et al. (2020) found that, under flooded conditions, soils not receiving straw emitted, on average, 38% less CH_4 than soils after straw incorporation, due to higher concentrations of dissolved organic matter (DOC) in the topsoil. Moreover, the authors tested a valuable mitigation option, consisting of collection and treatment of rice straw through anaerobic digestion (transforming the most labile C fractions into biogas) and returning the digestate to the paddy soil, to avoid C depletion of soil but maintaining its content in a stable form. A scenario of collection and selling of the rice straw was also assessed in Fusi et al. (2014), with a reduction of 52.5% of CH_4 emissions when compared to a baseline scenario in which straw was incorporated into the soil.

The survey carried out by Risobiosystems project showed that, under organic farming, rice cultivation in Europe follows different management strategies, which lead to different impacts on carbon footprint. The strategy of mulching from cover crop residues is highly innovative for rice crop in the area, and it brings several environmental benefits such as biodiversity and water quality protection. However, under long-term flooded conditions, cover crops represent a further C input, leading to large CH_4 paddy emissions. Organic farming is characterized by great yield variability, and yields are usually lower when compared to the conventional system (Bacenetti et al., 2016), mainly due to difficulties in weed management; and lower yields can cause a carbon footprint increase when referred to the product unit base. However, identification of the factors that determine yield variability and productivity gaps, as well as the use of genotypes with higher potential yield and higher competitiveness with the weeds, could help increasing rice productivity in organic farming (Orlando et al., 2020). Moreover, rotation systems typically practised in organic farming, in which rice cultivation is alternated to non-flooded crops, have the potential, on a multi-year basis, to decrease GHG emissions compared to conventional continuous system. The organic farming system also implies a different fertilization management, based on N supply with green manure, rotation with leguminous crops and organic fertilizers application; while mineral fertilizers in conventional systems cause large GHG emissions due to their industrial production and farm use, the use of organic fertilizers could have variable effects in terms of direct and indirect emissions. In the sample farms analysed in the Risobiosystems project, organic and mineral fertilizer production causes similar indirect GHG emissions, while direct soil emissions are differently affected. The type of organic fertilizer also influences the GHG emission: Bacenetti et al. (2016) found that the substitution of organic compost with cattle manure could reduce it.

Precision agriculture techniques applied in the sample farms of the Risobiosystems project also deal with variable fertilizer rate under conventional system; the application of this technology allowed us to increase rice yield and reduce GHG emissions per unit of product. Precision agriculture is increasingly considered as a powerful solution to mitigate the environmental impact of farming systems in Europe, improving the efficiency in input use and increasing farm profits. In the Italian rice system, Bacenetti et al. (2020), combining the use of remote sensing and ground estimates to derive maps of nitrogen nutrition index, achieved a reduction of 11.2% of GHG emissions compared with the baseline scenario due to yield increase.

Other mitigation options and carbon sequestration strategies that were adopted in some of the sample farms in the Risobiosystems project were agroforestry and minimum tillage. Agroforestry could improve carbon footprint performance also because of increase in total yields, due a double yield of trees and rice. Lehman et al. (2020) calculated land equivalent ratio (LER) values for agronomic production evaluation in several agroforestry systems in Europe; they found a range of 1.36–2.00, indicating that agroforestry systems were more productive by 36–100% compared to monocultures. In addition to environmental and economic benefits, agroforestry represents a carbon sequestration strategy for the carbon stock both in the biomass and in the soil, with variable mitigation potential, based on tree species growth rate, plant density and C input to soil through litter and roots.

Carbon sequestration potential of conservation agriculture practices, such as no or minimum tillage, could be higher than under conventional tillage. However, the increase in soil organic carbon may be due to the redistribution of carbon in the topsoil rather than a net increase in the overall soil carbon stock and largely depends on carbon input from the crops. Perego et al. (2019) found that in the paddy rice system in the Po Valley, the introduction of conservative practices also increased the economic efficiency, due to the reduction of tillage operations and related costs. However, the study concluded that the adoption of conservation agriculture practices is still in an initial phase in the area, and further the environmental and economic feasibility needs to be evaluated.

The anaerobic conditions induced by long flooding periods slow down the decomposition of organic matter, thus favouring soil organic carbon (SOC) storage. While rice paddy areas worldwide represent 9% of the global cropland, they accumulate 14% of total SOC pool in croplands, corresponding to 18 Pg SOC in the upper 100 cm of depth (Liu et al., 2021). There are agronomic practices that have proven to increase SOC stocks in rice paddies such as (i) addition of nitrogen and other nutrients, (ii) conservation agriculture practices, such as minimum or no-tillage, and (iii) returning rice straw into the soil (Chivenge et al., 2020).

Soil carbon sequestration practices deliver other environmental and agronomic benefits such as increase of soil fertility and soil biodiversity. However, CH_4 emissions can be greatly increased either during growing or flooded fallow seasons (Martínez-Eixarch et al., 2018; Martínez-Eixarch et al., 2021). Hence, the trade-off between GHG emissions and gains in SOC stock needs to be carefully examined. Soil C stock and net GWP have been studied in Spain by Belenguer-Manzanedo et al. (2021) in rice fields submitted to a set of four different post-harvest management options including the combination of timing of straw incorporation (early incorporation, in October, immediately after harvest vs late incorporation, in December) and flooding regime (winter flooding vs no-winter flooding). Changes in SOC stock were estimated by the net ecosystem carbon balance (NECB), which was calculated as the difference between C inputs (net primary production + straw addition) and C outputs (emissions of CO_2 and CH_4+ grain removal at harvest), while the net GWP, which estimates the global radiative forcing of the agroecosystem, was calculated as the difference between the GWP (cumulative CH_4 and N_2O emissions in CO_2-equivalent units) and changes in SOC stock. The results showed that rice fields acted as sink of carbon in all the treatments, with values ranging from 1.1 to 2.0 Mg C ha^{-1}, with the combination of no-winter flooding and late straw incorporation management that provided the largest C sink capacity, resulting from the less CH_4 and CO_2 emissions during fallow and growing seasons, respectively. However, water management determined the overall C balance in terms of radiative forcing, turning rice fields from sink in un-flooded to source of GHG emissions in flooded fields overwinter.

Conclusions and recommendations

The EU agricultural sector has had a limited contribution to GHG emission reduction until now, despite several measures that were taken to incentivize practices with expected positive impacts on Climate Change mitigation. However, new EU targets and policy initiatives such as the Green Deal and Farm to Fork strategies, as well as the increasing role of carbon sink as mitigation strategy set by the 2015 Paris Agreement and COP26 Glasgow conference, and the increasing perception of consumers of the impact of the agri-food sector, especially livestock, on GHG emissions will drive the agricultural sector towards reinforcing its role in the climate mitigation. Moreover, as climate change has direct negative effects on agriculture, for instance represented by the increase of extreme weather events and spread of new pests and diseases, farming systems are also looking for adaptation measures that could also give mitigation benefits. Adaptation and mitigation actions may also lead to further benefits to the farming systems, such as improvement of soil health, biodiversity and farm diversification.

In the case study reported, innovations adopted by the farmers are mainly related to organic agriculture practices, which have relevant environmental benefits but uncertain mitigation effects. While the reduction of herbicides and mineral fertilizers has clear positive effects on water quality and ecosystem protection, a low yield level due to poor weed control may lead to an increase of the carbon footprint per unit of product in the organic rice system. It is therefore necessary to improve agronomic strategies to close the yield gap and design rotation systems that have the potential, on a multi-year basis, to decreased GHG emissions compared to the conventional continuous system. The GHG emission mitigation strategies, such as water and straw management improvement, or carbon sequestration enhancement, such as conservative agriculture and agroforestry, can help in designing new rice farming systems with multiple benefits.

Note

1 Greenhouse gas emissions from land use, land use change and forestry.

References

Bacenetti, J., Fusi, A., Negri, M., Bocchi, S. and Fiala, M. (2016) 'Organic production systems: Sustainability assessment of rice in Italy', *Agriculture, Ecosystems & Environment*, vol. 225, pp. 33–44.
Bacenetti, J., Paleari, L., Tartarini, S., Vesely, F.M., Foi, M., Movedi, E., Ravasi, R.A, Bellopede, V., Durello, S., Ceravolo, C., Amicizia, F. and Confalonieri, R. (2020) 'May smart technologies reduce the environmental impact of nitrogen fertilization? A case study for paddy rice', *Science of The Total Environment*, vol. 715, p. 136956.
Belenguer-Manzanedo, M., Martínez-Eixarch, M., Fennessy, S., Camacho, A., Morant, D., Rochera, C., Picazo, A., Santamans, A.C., Miralles-Lorenzo, J., Camacho-Santamans, A. and Ibáñez, C (2021) 'Environmental and human drivers of carbon sequestration and greenhouse gas emissions in the Ebro Delta, Spain', In: Krauss, K.W., Zhu, Z., and Stagg, C.L. (eds.) *Wetland Carbon and Environmental Management*, pp.287–305. John Wiley and Sons, Hoboken, NJ.
Bertora, C., Cucu, M.A., Lerda, C., Peyron, M., Bardi, L., Gorra, R., Sacco, D., Celi, L. and Said-Pullicino, D. (2018) 'Dissolved organic carbon cycling, methane emissions and related microbial populations in temperate rice paddies with contrasting straw and water management', *Agriculture, Ecosystems & Environment*, vol. 265, pp. 292–306.
Bertora, C., Moretti, B., Peyron, M., Pelissetti, S., Lerda, C., Said-Pullicino, D., Milan, M., Fogliatto, S., Vidotto, F., Celi, L. and Sacco, D. (2020) 'Carbon input management in temperate rice paddies: Implications for methane emissions and crop response', *Italian Journal of Agronomy*, vol. 15(2), pp. 144–155.
Blengini, G.A. and Busto, M. (2009) 'The life cycle of rice: LCA of alternative agri-food chain management systems in Vercelli (Italy)', *Journal of Environmental Management*, vol. 90(3), pp. 1512–1522.
CFA (Cool Farm Alliance) (2019) 'CoolFarmTool'. Available at: https://coolfarmtool.org/
Coleman, K. and Jenkinson, D.S. (1996) 'RothC-26.3-A model for the turnover of carbon in soil', In: *Evaluation of soil organic matter models*, pp. 237–246. Springer, Berlin, Heidelberg.
Chivenge, P., Rubianes, F., Chin, D.V., Thach, T.V., Khang, V.T., Romasanta, R.R., Hung, N.V. and Trinh, M.V. (2020) 'Rice straw incorporation influences nutrient cycling and soil organic matter', In: *Sustainable rice straw management*, pp. 131–144. Springer, Cham.

COWI, Ecologic Institute and IEEP (2021) 'Technical Guidance Handbook—setting up and implementing result-based carbon farming mechanisms in the EU Report to the European Commission, DG Climate Action, under Contract No. CLIMA/C.3/ETU/2018/007'. COWI, Kongens Lyngby.

EC (European Commission) (2019a) 'Agridata'. Available at: https://agridata.ec.europa.eu/extensions/DataPortal/rice.html

EC (European Commission) (2019b) 'Communication on the European Green Deal, COM(2019)640'. Available at: https://eur-lex.europa.eu/resource.html?uri=cellar:b828d165-1c22-11ea-8c1f-01aa75ed71a1.0002.02/DOC_1&format=PDF

EC (European Commission) (2020) 'EU strategy to reduce methane emissions, COM (2020)663'. Available at: https://eur-lex.europa.eu/legal-content/EN/TXT/PDF/?uri=CELEX:52020DC0663&from=EN

EC (European Commission) (2021) 'Sustainable Carbon Cycles, COM(2021)800'. Available at: https://ec.europa.eu/clima/system/files/2021-12/com_2021_800_en_0.pdf

ECA (European Court of Auditors) (2021) 'Special report. Common agricultural policy and climate'. Available at: https://www.eca.europa.eu/Lists/ECADocuments/SR21_16/SR_CAP-and-Climate_EN.pdf

EEA (European Environmental Agency) (2019) 'EEA greenhouse gases—data viewer'. Available at: https://www.eea.europa.eu/data-and-maps/data/data-viewers/greenhouse-gases-viewer

ENRD (2016) 'Rural Development Programmes 2014-2020: Key facts & figures FOCUS AREA 5E: Carbon conservation /sequestration'. Available at: https://enrd.ec.europa.eu/sites/default/files/focus-area-summary_5e.pdf

Ente Nazionale Risi (2020) 'Evoluzione di mercato e sue prospettive—Relazione MIPAAF 2020'. Available at: http://www.enterisi.it/upload/enterisi/bilanci/RelazioneMIPAAF2020web_15916_2612.pdf

FAOSTAT (2019) 'Food and agricultural data'. Available at: https://www.fao.org/faostat/en/#home

Fusi, A., Bacenetti, J., González-García, S., Vercesi, A., Bocchi, S. and Fiala, M. (2014) 'Environmental profile of paddy rice cultivation with different straw management', *Science of the Total Environment*, vol. 494, pp. 119–128.

Hillier, J., Walter, C., Malin, D., Garcia-Suarez, T., Mila-i-Canals, L. and Smith, P. (2011) 'A farm-focused calculator for emissions from crop and livestock production', *Environmental Modelling & Software*, vol. 26(9), pp. 1070–1078.

Kraehmer, H., Thomas, C. and Vidotto, F. (2017) 'Rice production in Europe', In: *Rice Production Worldwide*, pp. 93–116. Springer, Cham.

Lehmann, L.M., Smith, J., Westaway, S., Pisanelli, A., Russo, G., Borek, R., Sandor, M., Gliga, A., Smith, L. and Ghaley, B.B. (2020) 'Productivity and economic evaluation of agroforestry systems for sustainable production of food and non-food products', *Sustainability*, vol. 12(13), p. 5429.

Leip, A., Carmona-Garcia, G. and Rossi, S. (2017) '*Mitigation Measures in the Agriculture, Forestry, and Other Land Use (AFOLU) Sector. Quantifying Mitigation Effects at the Farm Level and in National Greenhouse Gas Inventories*'. European Union, Luxembourg.

Liu, Y., Ge, T., van Groenigen, K.J., Yang, Y., Wang, P., Cheng, K., Zhu, Z., Wang, J., Li, Y., Guggenberger, G., Sardans, J., Penuelas, J., Wu, J. and Kuzyakov, Y. (2021) 'Rice paddy soils are a quantitatively important carbon store according to a global synthesis', *Communications Earth & Environment*, vol. 2(1), pp. 1–9.

Martineau, H., Wiltshire, J., Webb, J., Hart, K., Keenleyside, C., Baldock, D., Bell, H. and Watterson, J. (2016) '*Effective Performance of Tools for Climate Action Policy—Meta-review* of Common Agricultural Policy (CAP) Mainstreaming. Report for European Commission-DG Climate Action'. Ricardo-AEA, Didcot.

Martínez-Eixarch, M., Alcaraz, C., Viñas, M., Noguerol, J., Aranda, X., Prenafeta-Boldú, F.X., Saldaña-De la Vega, J.A., Català, M. and Ibáñez, C. (2018) 'Neglecting the fallow season can significantly underestimate annual methane emissions in Mediterranean rice fields', *PLoS One*, vol. 13(5), p. e0198081.

Martínez-Eixarch, M., Alcaraz, C., Guàrdia, M., Català-Forner, M., Bertomeu, A., Monaco, S., Cochrane, N., Oliver, V., Teh, Y.A. and Courtois, B. (2021) 'Multiple environmental benefits of alternate wetting and drying irrigation system with limited yield impact on European rice cultivation: the Ebre Delta case', *Agricultural Water Management*, vol. 258, p. 107164.

Matthews, A. (2012) 'Environmental public goods in the new CAP: Impact of greening proposals and possible alternatives'. European Union, Brussels.

Miniotti, E. F., Romani, M., Said-Pullicino, D., Facchi, A., Bertora, C., Peyron, M., Sacco, D., Bischetti, G. B., Lerda, C., Tenni, D., Gandolfi, C., and Celi, L. (2016) 'Agro-environmental sustainability of different water management practices in temperate rice agro-ecosystems', *Agriculture, Ecosystems & Environment*, vol. 222, pp. 235–248.

Monaco, S., Volante, A., Orasen, G., Cochrane, N., Oliver, V., Price, A. H., Teh, Y.A., Martínez-Eixarch, M., Thomas, C., Courtois, B. and Valé, G., (2021) 'Effects of the application of a moderate alternate wetting and drying technique on the performance of different European varieties in Northern Italy rice system', *Field Crops Research*, vol. 270, p. 108220.

Orlando, F., Alali, S., Vaglia, V., Pagliarino, E., Bacenetti, J. and Bocchi, S. (2020) 'Participatory approach for developing knowledge on organic rice farming: Management strategies and productive performance', *Agricultural systems*, vol. 178, p. 102739.

Perego, A., Rocca, A., Cattivelli, V., Tabaglio, V., Fiorini, A., Barbieri, S., Schillaci, C., Chiodini, M., E., Brenna, S. and Acutis, M. (2019) 'Agro-environmental aspects of conservation agriculture compared to conventional systems: A 3-year experience on 20 farms in the Po valley (Northern Italy)', *Agricultural Systems*, vol. 168, pp. 73–87.

Peyron, M., Bertora, C., Pelissetti, S., Said-Pullicino, D., Celi, L., Miniotti, E., Romani, M. and Sacco, D. (2016) 'Greenhouse gas emissions as affected by different water management practices in temperate rice paddies', *Agriculture, Ecosystems & Environment*, vol. 232, pp. 17–28.

RICA (2019) 'Area – Analisi dei Risultati Economici Aziendali'. Available at: https://arearica.crea.gov.it

Shirato, Y. and Yokozawa, M. (2005) 'Applying the Rothamsted carbon model for long-term experiments on Japanese paddy soils and modifying it by simple tuning of the decomposition rate', *Soil Science & Plant Nutrition*, vol. 51(3), pp. 405–415.

SINAB (2017) 'Sviluppo e trasferimento a sostegno della risicoltura biologica—Risobiosystems'. Available at: https://www.sinab.it/ricerca/sviluppo-e-trasferimento-sostegno-della-risicoltura-biologica-riso-biosystems

UNFCCC (2019) 'Greenhouse Gas Data'. Available at: https://unfccc.int/process-and-meetings#:0c4d2d14-7742-48fd-982e-d52b41b85bb0:adad877e-7830-4b8b-981a-5cdd34249c9d

Whittaker, C., McManus, M. C. and Smith, P. (2013) 'A comparison of carbon accounting tools for arable crops in the United Kingdom', *Environmental Modelling & Software*, vol. 46, pp. 228–239.

8 Agroecological farming approaches that enhance resilience and mitigation to climate change in vulnerable farming systems

Mehreteab Tesfai, Alamu Oladeji Emmanuel, Joyce Bakuwa Njoloma, Udaya Sekhar Nagothu, and Ngumayo Joel

Introduction

Current farming systems rely heavily on the intensive use of external resources and inputs such as water, mineral fertilizers, and pesticides to increase agricultural production (Bernard and Lux, 2017). Such farming systems have caused severe degradation of land water resources, soil depletion, increased outbreaks of pests and diseases, biodiversity loss, decline of ecosystem services (ESSs), and high levels of greenhouse gas emissions (e.g., FAOSTAT, 2020). There is a widespread recognition and growing concern that agricultural approaches based on *high-external inputs and resource-intensive farming systems* cannot deliver sustainable food and agricultural production (e.g., FAO, 2018) and it is likely that 'planetary boundaries' will even be further exceeded by such systems (e.g., Struik and Kuyper, 2017). Hence, more sustainable and affordable production methods are needed to protect and optimize the Earth's natural resources, while increasing productivity, adaptation, and mitigation to climate change. At the same time, the assumption is that sustainable agroecological farming systems provide several economic, environmental, social, and health benefits, and are the main prerequisite for food and nutrition security (e.g., Nguyen, 2018).

In recent years, key actors including regional governments, international agencies, civil society, and non-governmental organizations have demonstrated their commitments *to a new paradigm shift based on agroecology (AE)*. Some of these initiatives include (i) the new research and innovation programme by the European Commission (EC) (2020, 2021) 'Horizon Europe – Cluster 6: "Food, Bioeconomy, Natural resources, Agriculture and Environment"' launched in 2021 that supports a number of sub-priority topics on agroecology, and (ii) the Research Institute of Organic Agriculture (FiBL) and the FiBL project 'SysCom' in Kenya, Bolivia, and India (https://systems-comparison.fibl.org/). In addition, assessment reports, e.g., by IPCC (2019), FAO (2019), HLPE (2019), and UN Decade of Action on Nutrition (2016–2025), have all emphasized AE's potential

DOI: 10.4324/9781003273172-8

contributions to climate change mitigation and adaptation, biodiversity preservation, and ESSs. Other international and regional institutions and agencies like the AGRA (2016) and IPES-Food (2016), and international peasants' movement (e.g., La Via Campesina: https://viacampesina.org/en/international-peasants-voice/) are promoting AE as a potential to climate-neutral and resilient farming systems (CNRFSs).

Definition and concepts of agroecology

There has been continuous debate about the definition of agroecology, as evident from the literature (e.g., FAO, 2018; Wezel and Silva, 2017), with no widely accepted, common definition of agroecology yet. There are no clear, consensual boundaries between what is agroecological and what is not (HLPE, 2019). However, there is a consensus that agroecology embraces three dimensions: a transdisciplinary science, a set of principles and practices, and a social movement that is interlinked and complementary (Figure 8.1).

Agroecology is a powerful strategy that reduces the trade-offs between productivity and sustainability of agriculture and food systems (*social, economic, and environmental*) while ensuring 'no one is left behind' (Niggli et al., 2021). It promotes the diversity of crops and livestock, fields, farms, and landscapes, which altogether are key to improving the sustainability of food and agricultural systems, food actors' empowerment, and environmental health (von Braun et al., 2021). The agroecology approach has the potential to contribute to several Sustainable Development Goals (SDGs) of the United Nations as listed in Table 8.1.

The main *objectives of the chapter* are to (i) describe the common AE practices/approaches implemented in *agroforestry-based farming systems* and discuss their implications to ecological and socio-economic dimensions of AE; and (ii) recommend optimal combinations of AE practices/approaches that enhance food security, resilience, and mitigation to climate change in the different agroecological settings.

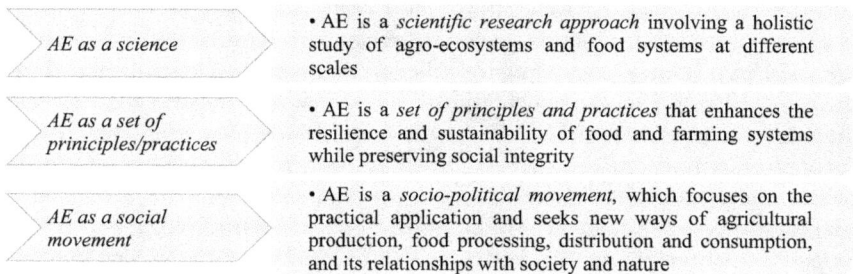

AE as a science	• AE is a *scientific research approach* involving a holistic study of agro-ecosystems and food systems at different scales
AE as a set of priniciples/practices	• AE is a *set of principles and practices* that enhances the resilience and sustainability of food and farming systems while preserving social integrity
AE as a social movement	• AE is a *socio-political movement*, which focuses on the practical application and seeks new ways of agricultural production, food processing, distribution and consumption, and its relationships with society and nature

Figure 8.1 Three dimensions of agroecology.
Source: Authors' own compilation.

Table 8.1 Potential AE contributions to relevant SDGs and specific targets along with references

SDGs	Targets*	References (examples)
SDG 1: *no poverty*	1.3, 1.4, 1.5	Niggli et al. (2021)
SDG 2: *zero hunger*	2.1, 2.2, 2.3, 2.4, 2.5	Deaconu et al. (2021)
SDG 3: *good health and well-being*	3.9	FAO (2018)
SDG 5: *gender equality*	5.1	von Braun et al. (2021)
SDG 6: *clean water and sanitation*	6.4	FAO (2018)
SDG 8: *decent work and economic growth*	8.3, 8.5	CNS-FAO (2021); ILO (2018)
SDG 10: *reduce inequalities*	10.2	FAO (2018)
SDG 12: *sustainable consumption and production*	12.1, 12.2, 12.3	FAO (2018)
SDG 13: *climate action*	13.1, 13.2, 13.3	Leippert et al. (2020)
SDG 15: *life on land*	15.1	Altieri and Nicholls (2018)
SDG 16: *peace, justice, and strong institutions*	16.7	FAO (2018)
SDG 17: *partnerships for development goals*	17.6, 17.9	FAO (2018)

Source: Authors' own compilation.
* Note that the specific targets and corresponding indicator descriptions can be found at https://unstats.un.org/sdgs/indicators/indicators-list/.

Elements of agroecology and their implications for practice

Table 8.2 provides the ten elements of the AE framework as defined by FAO (2019), and their implications for practice. According to the report, the ten AE elements encompass ecological characteristics of AE systems (*diversity, synergies, efficiency, resilience, and recycling*), social characteristics (*co-creation and sharing of knowledge, human and social values, culture and food traditions*), and the enabling political and economic environment (*responsible governance, circular and solidarity economy*). Most of these elements of AE relate well to the 13 principles of AE developed by HLPE (2019).

The above-mentioned ten elements of AE are interconnected and interdependent with one another. These elements of AE and their practices also fit well with the aims and goals of the regenerative agriculture (e.g., *produce more from less*: less land area, less input of chemicals, less use of water, less emission of greenhouse gases, less risk of soil degradation, and less use of energy-based inputs). However, a detailed discussion on regenerative agriculture is beyond the scope of this chapter.

Main barriers for agroecology adoption

The multiple benefits of AE have been demonstrated in specific contexts and gained prominence in scientific literature, and agricultural and political discourse. Despite this, AE has not been mainstreamed and not widely adopted in

Table 8.2 The ten elements of AE (FAO, 2019) and their implications when put in
          practice

AE elements	Implications for practice
• Diversity	Producing and consuming a *diverse range of cereals, pulses, fruits, vegetables, and animal-source* products contributes to improved nutritional outcomes, diversity in diets and markets
• Synergies	Combining annual and perennial crops, livestock and aquatic farming, trees, soils, water, and others enhances synergies in the context of climate change
• Efficiency	Producers are able to *use fewer external resources, reduce costs,* and *reduce environmental impacts* by enhancing bio-based measures, biological control, recycling biomass, nutrients, and water
• Recycling	AE practices support biological processes by imitating natural ecosystems that drive recycling of nutrients, biomass, and water to minimize waste and pollution
• Resilience	Diversified AE systems are more resilient and have *greater capacity to recover from* extreme weather events (e.g., *drought, floods*) and better resistance to *pest and disease attack*
• Co-creation/sharing of knowledge	Blending *indigenous knowledge, practical knowledge, and scientific knowledge* addresses challenges across food systems and resilience to climate change
• Human and social values	AE places a strong emphasis on human *dignity, equity, inclusion, and environmental justice* for all and thereby contributing to improved livelihoods
• Culture and food traditions	AE plays an important role in re-balancing *traditional and modern food habits* by promoting healthy food production and consumption, and by supporting food sovereignty (the right to adequate food)
• Responsible governance	*Transparent, accountable, and inclusive* governance mechanisms support producers to transform their farming systems following the AE practices
• Circular and solidarity economy	AE seeks to *reconnect producers and consumers*, prioritizes local markets, and supports economic development by creating short circular value chains that reduce food losses and wastes

Source: Author's own elaboration adapted from FAO (2019).

different farming systems and agroecological zones (AEZs) worldwide. The main
barriers to widespread adoption and upscaling of AE practices/approaches at field/
farm/landscape levels include the following:

i *Lack of awareness and knowledge about AE:* Despite successful AE experiences
   in some regions of the world (CNS-FAO, 2021), there is a lack of awareness among key stakeholders (e.g., decision-makers) and the public on the
   potential of AE to tackle environmental, social, and economic challenges
   posed by climate change and its contributions to achieving multiple targets
   of the SDGs (see Table 8.1). Moreover, limited information is available on
   the extent to which AE can be applied to larger farms (Parmentier, 2014) and

the economic and social impacts of AE for different groups in the farming communities (Bezner Kerr et al., 2019). There are differences in approaches and ideologies resulting in conflicts of interest between proponents and opponents of AE. For example, ideological differences exist among scientific communities and fertilizer companies (e.g., Yara International ASA) that promote increased external farm inputs (e.g., chemical fertilizers). Such conflicts need to be resolved by minimizing the trade-offs while maximizing synergies and complementarities of ecological and socio-economic dimensions of AE (Mockshell and Kamanda, 2018).

ii *Insufficient investments on AE research and extension systems*: Current research and extension systems do not sufficiently address the key AE principles/ practices when compared to the investments on conventional agriculture (IAASTD, 2009). There is a lack of incentive in long-term research and limited funding available, e.g., to assess the yield gap between 'intensive farming systems' and 'AE systems' (CNS-FAO, 2021). The current agricultural research and extension systems predominantly focus on single disciplines, single technology, and single commodity, and use top-down extension models to transfer knowledge/technology.

iii *Additional labour costs*: Adoption of AE farming practices such as AC systems with agroforestry incurs additional labour cost (Schoonhiven and Runhaar, 2018). This is a challenge for smallholder farmers who cannot afford especially in the initial year of establishment. Therefore, farmers are not motivated to adopt unless the immediate net benefits or profits of AE farming are visible.

iv *Inadequate policy support, gender integration, multi-actor partnerships*: Lack of policy and institutional enabling environments deters widespread implementation of AE (Anderson et al., 2020). Gender integration in AE projects plays a crucial role in adopting AE practices, for instance, in the African context (Bezner Kerr et al., 2019). However, much has not been done at the policy or practice level to strengthen gender integration in promoting agroecology. There is a lack of coordination/collaboration among stakeholders in implementing AE projects (Ayala-Orozco et al., 2018), probably due to the absence of relevant platforms/networks/partnerships.

v *Lack of evidence on the interactive effects across AE practices*: Despite the extensive literature on AE farming systems in the form of scientific and popular publications, there is a lack of evidence on the *optimal combinations of integrated AE practices and their impacts* in different agroecological regions, farming context and scales.

There is no 'one-size-fits-all' solution to AE farming system challenges (Schader et al., 2014) but to use a combination of solutions. Combining sustainable intensification (SI) approaches with AE approach/practices is the way forward to address the multiple challenges faced by smallholders. The key principles of SI (Box 8.1) are in line with the principles of AE elements. Both SI and AE have a common objective, i.e., to achieve food and nutrition security (FNS) while reducing

---

Box 8.1 Components of sustainable intensification

i *Increasing production, income, nutrition, or other returns* on the same amount of, or less land and water by efficient and prudent use of inputs and productive use of knowledge and capacity to adapt, innovate, and scale up.

ii *Minimizing greenhouse gas emissions by* increasing natural capital and the flow of environmental services, and reducing impact on forests through alternative energy sources.

iii *Strengthening resilience and reducing environmental impacts by* adopting innovative technologies and processes while minimizing inputs that have adverse impacts on people and environment.

Source: Pretty et al. (2011)

---

negative impacts on the environment (Bernard et al., 2017). They can generate healthy soils, crops, and animals, which is a core element of regenerative agriculture (Newton et al., 2020). One basic divergence between the two is SI focuses on increasing the food production side of the food systems. At the same time, AE addresses the whole food systems along their value chains and relationships with society and nature (Lampkin et al., 2015). In this chapter, we will focus on the convergence of SI and AE by promoting the essential elements of AE (Table 8.2) and their implications to practice, science, and policy.

The chapter has been divided into four main sections. The first section introduces AE definition and concepts, principles and practices, the potential for sustainable developments, gaps and barriers for adoption and upscaling. This is followed by case study descriptions of farmer-led AC with *Gliricidia* agroforestry demonstration trials in Zambia and the methodological approaches used. Then, a detailed analysis of the research results is presented and discussed including fundamental AE principles, practices, and policy. Towards the end, optimal combinations of the AE practices that enhance food security, resilience, and mitigation to climate change are recommended.

Case study and methods used

This chapter presents some findings from a multi-disciplinary alley cropping (AC)-*Gliricidia* agroforestry project[1] in Zambia, as a case study. Alley cropping (also sometimes referred to as 'Hedgerow intercropping') is defined as the practice of planting rows of trees and/or shrubs to create alleys with companion crops in between. The AC with *Gliricidia* agroforestry system combines maize and legumes (groundnuts and soybeans), where smallholder farmers have implemented a set of AE principles/practices since 2019. The study's primary objectives were to (i) monitor the soil nutrients, in particular nitrogen and organic carbon inputs and outputs, under agroforestry systems (e.g., *Gliricidia sepium*) at different levels of intensification and measure crop yields and (ii) assess the impact of agroforestry-based interventions on the nutrients of selected crops to address whether agroforestry practices result in healthier and nutrient-rich food crops.

The study was conducted in maize growing districts of eastern Zambia in five selected *Chiefdoms* (an area/region governed by a chief). The *Chiefdoms* covering the study are *Mkanda* (Chipangali district), *Zumwanda* (Lundazi district), *Mwasemphangwe and Chikomeni* (Lumezi district), and *Magodi* (Chasefu district), as shown in Figure 8.2 and Table 8.3.

The main farming system in the study areas is maize-based monocropping with low-input and low-output smallholder agriculture (Table 8.3). Farmers (including women) lack access to good quality seeds and adaptive knowledge about climate-resilient crops, crop residues, and soil management practices, among others. Hence, gender integration becomes a challenge in the overall context of smallholder agriculture. Crop diversification with legumes and/or agroforestry in particular AC systems will provide multiple benefits to smallholder farmers who are vulnerable to climate change. Alley cropping (of maize-legumes that includes groundnuts and soybeans) with agroforestry trees such as *Gliricidia sepium* was recently introduced in the study areas.

In the case study areas, soil fertility is declining over time due to several factors, among others burning of crop residues, leading to low organic matter levels of the soils. Soil health/soil quality, defined as *'the continued capacity of soils to function*

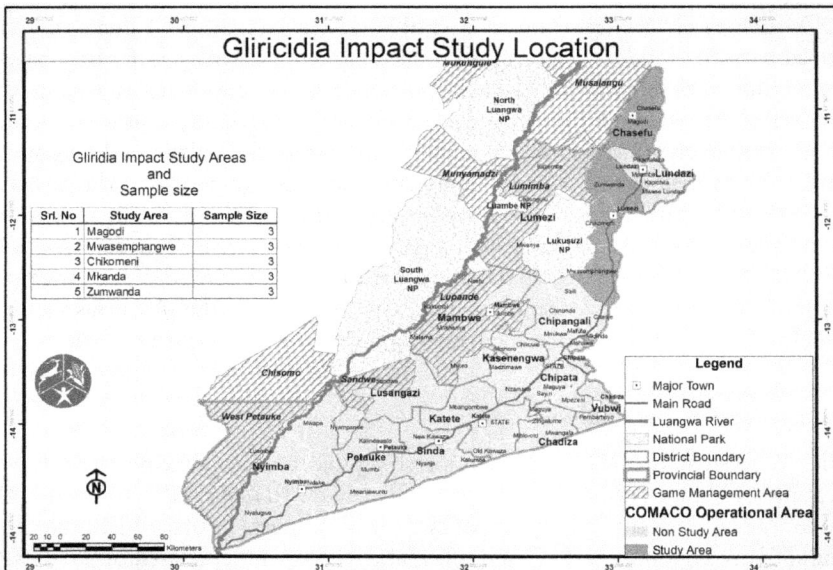

Figure 8.2 Map of the case study sites in the eastern Zambia.

Table 8.3 Summary of the general characteristics of the study sites in eastern Zambia

Features of study sites	
Agro-climatic conditions	Tropical Savanna
Elevation (above sea level)	1,140–1,143 meter
Precipitation (range)	923–1,023 mm/yr
Air temperature (range)	18–27°C
Soil types (dominant)	Red-brownish clayey to loamy soils
Farming systems	Maize-based monocropping under rainfed
Major crops	Maize, groundnut, beans, cotton, sunflower, tobacco
On-farm tree (dominant)	*Gliricidia sepium*
Livestock	Chicken, cattle, goats, pigs
Ecological constraints	Low soil fertility, erratic rainfall/dry spells, crop residues burning
Socio-economic constraints	Small land sizes, poverty, food/nutrition insecurity, high population pressure, lack of access to quality seeds
Opportunities explored	Farm diversification with maize, legumes, and alley cropping

Source: Authors' own compilation.

properly and provide the required ecosystem services and goods', is essential for improving crop yield and crop nutritional quality. The crop nutritional quality largely depends on the composition and concentration of nutrients available in the soil. Maintaining healthy soil ensures nutritious, tasty, and safe foods, and enhances resilience and mitigation to climate change which are essential for achieving the SDGs such as SDG 2 (*zero hunger*) and SDG 13 (*climate action*). Hence, there is a need to understand whether crops produced under AC with agroforestry-based systems are more nutritious than those produced in conventional systems.

A set of AE practices that include *AC of maize, groundnuts, soybean with Gliricidia, conservation agriculture, composting/leaf manuring, residue mulching* were implemented in selected on-farm demonstration trials ($n = 15$) in the eastern province of Zambia through a farmer-led approach. Farmer-to-farmer extension services backed these demo trials through farmer field days for broader adoption of *Gliricidia* and knowledge sharing. Farmer-to-farmer knowledge exchange on AE farming practices was also carried out through multimedia platforms such as weekly radio broadcasts to the farming community in the case study areas.

Data collection and analysis

The AC system with *Gliricidia* agroforestry project in Zambia involved seven treatment (T) plots, i.e., intercropping of *Gliricidia* with maize (T1), soybean (T4), groundnuts (T5) and sole cropping of maize with mineral fertilizer (T2), sole soybean (T6), sole groundnut (T7) and sole maize with no mineral fertilizer (T3) used as a control. Soil samples ($n = 178$) were collected from three random positions in each treatment plot of the 15 on-farm demo trials. The soils were recovered from the topsoil and subsoil layer using a soil auger and were analysed for selected soil chemical and physical properties such as soil bulk density that was used in the computation of estimating carbon stocks in the soils.

Crop samples of maize, soybean, and groundnuts ($n = 88$) were collected using the standard sampling protocol developed for the project. The crop samples were cleaned, subsampled, and milled to a 0.5 mm particle size. The milled samples were evaluated for the nutritional contents (fat, ash, protein, starch, crude fibre, sugar, amylose, and total carbohydrate) and antinutritional contents (phytate and tannin) using standard laboratory methods of analysis of the Association of Analytical Chemist International (AOAC). The data generated were analysed for descriptive statistics and analysis of variance (ANOVA) using Statistical Analysis Software (SAS) version 9.4. The *F*-test was used for statistical significance. The treatment means were compared using the least significant difference (LSD) tests at $P < 0.05$.

Results and discussion

In the following subsections, the effects of AC systems with *Gliricidia* agroforestry interventions on soil health, crop yield, crop nutrient quality, and climate change adaptation and mitigation are presented and discussed.

Soil health assessment

The status of soil health was assessed through a set of measurable physical, chemical, and biological indicators. These include soil organic carbon (SOC) and/or organic matter (OM), soil nutrients (in particular nitrogen), soil pH (acidity), and soil structure related to bulk density of soils, used as a reflection of overall soil health indicators.

Over the two growing seasons (2019/2020 and 2020/2021), the SOC contents of the sampled soils ($n = 182$) varied in the range of 0.32% (0.60% organic matter, OM) to 1.10% (2.0% OM), which is very low to low despite some positive increase observed since the incorporation of *Gliricidia* leaf manure (Figure 8.3) into the treated soils (T1, T4, T5). The soils in the study sites would benefit from additions of organic fertilizers obtained from *Gliricidia* leaf manure and nitrogen-fixing legumes. However, retention and accumulation of OM and SOC storage in the soil require considerable time. Thus, repeated application of diverse organic sources (such as leaf biomass and crop residues) will stimulate microbial community growth and sequestration of carbon in the soils (Moebius-Clune et al., 2016).

The mean carbon stock per treatment ranged from 17.6 to 25.6 C t/ha (Figure 8.4) and similar results have been reported in different farming systems. For example, agri-silviculture agroforestry systems could store about 27 C t/ha and rainfed crop production systems in semi-arid areas about 16 C t/ha. The highest carbon stock was measured in T3 (sole Maize + no Mineral fertilization). The possible explanation for this could be that retention and accumulation of OM/OC storage in the soil requires a considerable time. Repeated application of diverse

Figure 8.3 Woman farmer incorporating *Gliricidia* tree leaves into the soils.

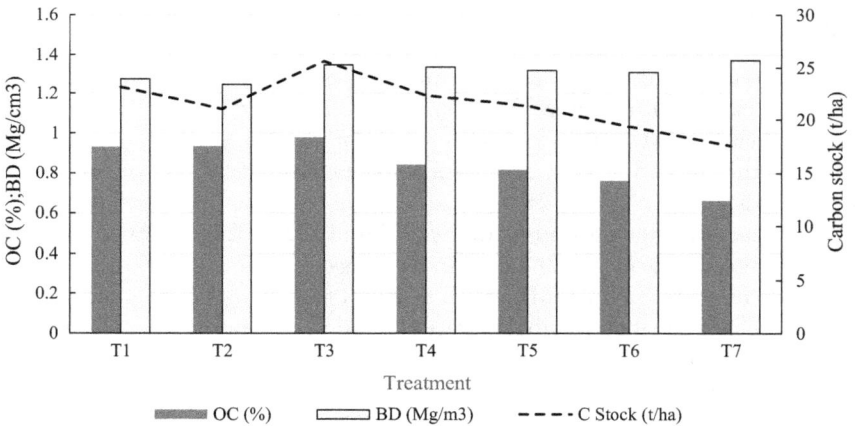

Figure 8.4 Mean organic carbon, BD, and carbon stock estimates from the seven treatment plots: T1: *Gliricidia* + Maize intercrop, T2: sole Maize + Mineral fertilization, T3: sole Maize + no Mineral fertilization, T4: *Gliricidia* + Soybean, T5: *Gliricidia* + Groundnuts, T6: sole Soybean, and T7: sole Groundnuts.
Source: Authors' own analysis.

organic sources (such as green manures and crop residues) in the long term will stimulate both microbial community growth and the stabilization (sequestration) of carbon in aggregates (Moebius-Clune et al., 2016). The magnitude of changes in soil OM depends on the quantity and quality of prunings, pedo-climatic conditions, and the system management as a whole (Makumba et al., 2007). There is

Table 8.4 Average grain yields of maize, groundnut, and soybean by treatment ($n = 15$ demo plots)

Treatment	Maize (kg/ha)		Soybean (kg/ha)		Groundnut (kg/ha)	
	2019/2020	2020/2021	2019/2020	2020/2021	2019/2020	2020/2021
T1	819.3	4520	–	–	–	–
T2	820.3	5954	–	–	–	–
T3	540.1	1227	–	–	–	–
T4	–	–	329	910	–	–
T5	–	–	–	–	393.2	708
T6	–	–	328.7	825	–	–
T7	–	–	–	–	372.4	737

T1: *Gliricidia* + Maize intercrop, T2: sole Maize + Mineral fertilization, T3: sole Maize + no Mineral fertilization, T4: *Gliricidia* + Soybean, T5: *Gliricidia* + Groundnuts, T6: sole Soybean, and T7: sole Groundnuts.
Source: Authors' own analysis.

a great potential to mitigate climate change through AC systems with *Gliricidia* agroforestry interventions in the study areas. Selling carbon credits may provide another source of income for farmers, but policies need to be in place for encouraging carbon markets to benefit smallholders practising AE.

The bulk density (BD) of the soils ranged between 1.26 g/cm^3 and 1.36 g/cm^3, which is within the range of 1.0–1.7 g/cm^3 for typical agricultural soils (Brady and Weil, 2002). The soil bulk density serves as an indicator of compaction, root growth, and water movement in the soils, and it is a good indicator for soil health.

Crop yield assessment

Table 8.4 presents average grain yields (kg/ha) of maize, groundnut, and soybean by treatments. *Gliricidia*-maize intercrop (T1) and maize with 100% mineral fertilizer (T2) did not differ significantly ($P > 0.05$). In the 2020/2021 season, the yields (maize variety MZ521) are within the potential national yield range: 4,-500–6,000 kg/ha. The maize yield from the control plots (T3) rendered much lower compared with T1 and T2. There was an increase in yields in the second season for the groundnuts and soybeans with no significant differences between treatments. In all treated crops, the grain yields increased by more than two to three folds in 2020/2021. The yield increase can be attributed to good field management practices, including *Gliricidia* leaf biomass incorporation, conservation agriculture practices, and crop diversification.

Crop nutrient assessment

Table 8.5 shows the nutritional properties (NPs) and antinutritional properties (ANPs) of maize samples by treatment. The treatment showed a significant impact ($P < 0.05$) on all NPs and ANPs except for ash (inorganic matter) which was not significant at $P > 0.05$. The non-significant difference with ash content of the

Table 8.5 Nutritional and antinutritional properties of maize by treatment (*n* = 37)

Properties	T1 (Gliricidia + Maize intercrop)		T2 (sole Maize + Mineral fertilizer)		T3 (sole Maize + no Mineral fertilizer)		Pr > F (T)	Pr > F (F × T)
	Mean	SD	Mean	SD	Mean	SD		
MC, %	6.64 b	0.93	6.31 b	0.67	7.09 a	0.88	<0.0001	<0.0001
Ash, %	1.31 a	0.09	1.30 a	0.07	1.27 a	0.08	0.0927	0.0048
Fat, %	4.72 b	1.02	5.66 a	1.12	5.16 b	1.27	<0.0001	<0.0001
Protein, %	6.28 b	0.98	7.14 a	1.22	5.93 c	0.93	<0.0001	<0.0001
CF, %	4.00 b	1.28	4.06 b	1.26	4.41 a	1.49	<0.0001	<0.0001
Sugar, %	2.76 ab	0.8	2.63 b	0.72	3.01 a	0.79	0.0079	0.0130
Starch, %	71.89 b	1.05	72.42 a	1.12	72.55 a	1.19	<0.0001	<0.0001
Amylose, %	28.35 b	1.51	28.80 ab	2.48	29.55 a	2.6	0.0434	0.1604
CHO, %	77.05 a	2.65	75.53 c	1.96	76.14 b	2.19	<0.0001	<0.0001
Phytic acid, %	2.27 ab	1.97	1.98 b	1.71	2.52 a	2.18	0.0034	<0.0001
Tannin, mg/g	3.10 c	0.72	3.42 b	0.95	3.62 a	0.6	<0.0001	<0.0001

MC = moisture content; CF = crude fibre; CHO = total carbohydrate; SD: standard deviation, F: F statistic, T: test statistic. Mean values with different letters in the same row are significantly different at *P* <0.05. *Pr > F*: this is the *P*-value associated with the F statistic of a given effect and test statistic. Source: Authors' own analysis from field data.

maize samples agrees with the study by Ogunyemi et al. (2018), who also reported no significant difference in the ash content of maize samples subjected to different treatments (NPK and biochar fertilized). The mean values for ash, fat, and protein contents obtained for the *Gliricidia*-Maize intercrop (T1) without mineral fertilizer are higher than the results reported by Ogunyemi et al. (2018) for maize using biochar fertilizer. It implies that *Gliricidia* has a better effect on nutritional properties than mineral fertilizer (NPK) and biochar. Also, *Gliricidia*-Maize intercrop (T1) without mineral fertilizer significantly reduced the tannin and phytic contents of maize samples compared with the control (T3).

Table 8.6 presents treatment effects on the soybean samples' NPs and ANPs. Both treatments had a significant (*P* < 0.05) effect on fat, amylose, total carbohydrate (CHO), energy, and ANPs. The result agrees with the studies by Etiosa et al. (2017) and Alamu et al. (2019), who reported similar values for soya bean seeds. A higher mean value was observed for protein, starch, amylose, crude fibre, and CHO contents when *Gliricidia* + soybean intercrop (T4) was used. There were lower mean values for ANPs in T4 but comparable ash contents and significantly lower fat contents (*P* < 0.05). The observation is similar to what Alamu et al. (2019) reported, where they observed low values of ANPs for the soybean samples taken from integrated soil management practices plots. The amylose and CHO contents were significantly increased while tannin and phytic acid contents were reduced in T4. Some soybean samples from T4 showed higher ash, protein, and carbohydrate contents but lower phytic acid and tannin contents than farmer plots from T6 (sole soybean).

Table 8.6 Nutritional and antinutritional properties of soybean by treatment (n = 26)

Parameters	T4 (Gliricidia + Soybean)		T6 (sole Soybean)		Pr > F(T)	Pr > F (F × T)
	Mean	SD	Mean	SD		
MC, %	8.08 a	1.08	8.10 a	1.08	0.9763	0.0002
Ash, %	5.55 a	0.83	5.81 a	0.61	0.1080	0.0127
Fat, %	18.65 b	3.59	20.98 a	5.07	<0.0001	<0.0001
Protein, %	37.73 a	3.20	36.98 a	4.04	0.1065	<0.0001
Sugar, %	5.45 a	0.75	5.62 a	0.69	0.0786	0.0126
Starch, %	22.86 a	0.94	22.95 a	0.96	0.9306	0.0006
Amylose, %	1.76 a	0.51	1.64 b	0.30	0.0970	0.0049
CF, %	2.11 a	0.32	2.04 a	0.37	0.4271	0.7015
CHO, %	27.88 a	6.06	26.09 b	6.15	0.0134	<0.0001
Phytic acid, %	6.47 b	1.14	7.09 a	1.19	0.0758	0.1953
Tannin, mg/g	3.88 b	0.92	5.02 a	1.70	0.0006	0.0192

Source: Authors' own analysis from field data.

Table 8.7 Nutritional and antinutritional properties of groundnut by treatment (n = 25)

Properties	T5 (Gliricidia + Groundnuts)		T7 (sole Groundnuts)		Pr > F (F)	Pr > F (F × T)
	Mean	SD	Mean	SD		
MC, %	5.69 a	0.84	5.65 a	0.52	0.4305	<0.0001
Ash, %	2.57 a	0.15	2.60 a	0.16	0.0925	<0.0001
Fat, %	47.23 a	6.72	44.97 b	6.46	<0.0001	<0.0001
Protein, %	19.01 a	2.28	17.94 b	2.21	<0.0001	<0.0001
Sugar, %	4.05 b	0.61	4.15 a	0.66	<0.0001	<0.0001
Starch, %	23.77 b	1.08	24.08 a	0.92	0.0011	<0.0001
Amylose, %	1.66 a	0.50	1.68 a	0.59	0.4802	<0.0001
CF, %	3.29 b	0.83	3.44 a	0.92	<0.0001	<0.0001
CHO, %	22.22 b	7.27	25.42 a	7.23	<0.0001	<0.0001
Phytic acid, %	4.38 a	0.84	4.35 a	0.75	0.8221	<0.0001
Tannin, mg/g	6.44 a	1.90	6.09 b	2.02	0.0490	0.0008

Source: Authors' own analysis from field data.

Table 8.7 shows the mean values and treatment effects on NPs and ANPs of groundnut. The result for nutritional properties of groundnut reported agrees with previously published studies on the proximate composition of groundnut samples (Asibuo et al., 2008; Atasie et al., 2009). Both treatments (T5 and T7) exhibited a significant effect ($P < 0.05$) on fat, protein, sugar, starch, crude fibre (CF), total carbohydrate (CHO), total energy, and tannin content of groundnut, but a non-significant effect on ash, amylose, and phytic acid at $P > 0.05$.

The mean values of crop samples from *Gliricidia* + Groundnut (T5) were higher in fat, protein, tannin, and bulk density but lower in starch, CF, and CHO than with sole Groundnuts (T7). The implication is that T5 significantly increased the crop's fat, protein, and tannin levels. Goudiaby et al. (2020) reported a non-significant effect of groundnut intercropped with *Eucalyptus camaldulensis* tree on the proximate content of the crop except for the grain yield. This implies that the *Gliricidia*-groundnut intercropping improved nutritional properties of groundnuts compared to treatment using *E. camaldulensis*.

It can be summarized that the *Gliricidia* + Maize intercrop (i.e., T1) showed the highest mean value of ash, fat, protein, and total carbohydrate (CHO) contents. A higher mean value of protein, starch, amylose, crude fibre, and CHO contents was measured in *Gliricidia* + Soybean intercrop (T4). *Gliricidia* + Groundnut intercrop (T5) significantly increased the fat and protein contents of groundnuts. *Gliricidia* + Maize intercrop (T1) significantly ($P < 0.05$) reduced the tannin and phytic contents of maize samples compared to the control (sole Maize: T3). Lower mean values of tannin and phytic acid were observed in T4 than sole Soybean (T6). A lower value of phytic acid but increased tannin level was measured in T5. Thus, intercropping with the *Gliricidia* improves the nutritional quality of maize, soybean, and groundnut and decreases the antinutritional qualities of the legumes.

Optimal combinations of AE practices/approaches

Table 8.8 presents the different AE practices/approaches that have been implemented in the case study sites of eastern Zambia (AC systems *with Gliricidia agroforestry*). The AE practices/approaches addressed more than one element/principle of AE (see Table 8.2) and contributed to enhancing the sustainability of AE farming from the point of view of ecological, social, and economic dimensions.

For instance, crop diversification through intercropping of cereals with legumes and AC with *Gliricidia* trees enhanced diversity (in crops, trees, habitat, food diets, markets), synergies (combining annual and perennial plants), resilience (to climate change), and culture and food traditions (increasing healthy food production and consumption and supporting the right to adequate food).

Table 8.8 Matching the AE practices/approaches implemented in the case study sites with the most appropriate AE elements/principles

Matching	AE practices/approaches	AE elements/principles
a), b), e)	Crop rotation with legumes	a) Diversity
a), b), e), h)	Intercropping with legumes	b) Synergies
b), c), e)	Conservation agriculture	c) Efficiency
a), b), e), h)	Alley cropping with *Gliricidia*	d) Recycling
c), d)	Composting, leaf manuring	e) Resilience
c), d)	Residue mulching	f) Co-creation and sharing of knowledge
a), e)	Agrobiodiversity	g) Human and social values
f)	Multi-media platforms	h) Culture and food traditions
b), f), g), i)	Stakeholder engagement	i) Governance (responsible/effective)
d), g), j)	AE products value chain	j) Circular and solidarity economy

Source: Author's own analysis.

Climate change adaptation and mitigation

In the *Gliricidia*-treated plots (i.e., T1, T4, and T5), the main sources of addition of nitrogen into the soils were the incorporation of leaf biomass from *Gliricidia* trees, atmospheric N-fixation by legumes (in this case, groundnuts, soybeans, and *Gliricidia*), and atmospheric deposition by rain. Total N inputs from these organic sources were estimated at 468–500 kg N/ha (data not shown). However, only a small proportion of this organic N becomes plant-available during a growing season (Horneck et al., 2011). The remaining part of the organic N will be mineralized and made available for the succeeding crops.

However, the AC systems with *Gliricidia* have improved the soils' organic matter content and increased carbon stocks (Figure 8.4). This will reduce the need for nitrogen-based fertilizers, which contribute to mitigating nitrous oxide (N_2O–N) emissions, a potent green-house gas. The average carbon sequestration in the soils was about 22 C t/ha (Figure 8.4). This implies that about 81 CO_2 equiv. t/ha was prevented from being released to the atmosphere, considering 1-tonne organic carbon reduces about 3.7-tonne atmospheric CO_2 equiv. In addition, reduced/no-tillage practices using animal-drawn rippers and hand seeding will also minimize carbon dioxide emissions in the long term. The AE practices such as intercropping legumes with *Gliricidia* agroforestry and soil mulching with residues can increase climate resilience to drought and dry spells.

In the following subsections, a brief discussion is given on how the AC systems with *Gliricidia* agroforestry project have addressed the key elements and principles of AE.

Addressing the ecological dimensions of agroecology

i *Diversity:* Regarding crop/tree and food diversity, farmers planted *Gliricidia sepium* seedlings in between maize, soybean, and groundnut fields (as AC) for food and sale and soil fertility improvements. Farmers in the project have diversified crop produce, of cereals (maize) and pulses (soybeans and groundnuts), contributing to diet diversity and improved nutrition. Multipurpose leguminous trees such as *Gliricidia* are used for improving soil fertility, reducing soil erosion, controlling striga weed, providing fuelwood (including charcoal), and forage for honey production. *Gliricidia* leaves are rich in crude protein (>20%) and highly digestible, and low in fibre and tannin contents, making it good fodder for livestock (refer Tables 8.5–8.7).

ii *Synergies:* The demonstration trials on AC with *Gliricidia* trees enhanced synergies of resource use such as nutrients. For instance, the maize plants received nitrogen from the nitrogen fixed by soybeans and/or groundnuts and decomposed leaf biomass of *Gliricidia* tree. Synergistic interactions between annual crops (maize, soybeans, and groundnuts) and the leguminous agroforestry trees (*Gliricidia*) enhance both soil and crop productivity resulting in increased crop yields. However, trade-offs such as competition for light in AC systems with agroforestry trees (e.g., *Gliricidia*) could be minimized by adopting good agronomic practices such as seedbed preparation, early planting, and weed management (Sida et al., 2018).

iii   *Efficiency:* Incorporation of *Gliricidia* leaf biomass improves resource use efficiency. *Gliricidia* trees produce large quantities of leaf biomass and contribute to increased soil productivity and crop yields over time. The decomposing *Gliricidia* leaf biomass enriches the soils with macronutrients such as nitrogen that support crop growth. This eventually leads to reduced external inputs of chemical fertilizers. *Gliricidia* trees produce high-quality leaf biomass that contains as much as 4% total N in their leaves.

Implementing AC that consisting of maize, different legumes, and trees effectively contributes to improving land use efficiency where land equivalent ratios are greater than >1. This indicates AC practices are more productive in the use of land resources where landholding size is shrinking, e.g., in the case of Zambia.

iv   *Recycling: Gliricidia sepium* is a fast-growing leguminous agroforestry tree with relatively deep root system that captures leached nutrients along the soil profile. Thus, nutrients accumulated in layers below the root zone of annual crops can be accessed. These nutrients absorbed by the root system of the trees become inputs when transferred to the soil surface in the form of litter and other plant residues. Incorporating nutrient-rich tree leaves, especially leaves of leguminous trees like *Gliricidia*, can be considered as a potential solution towards improving soil fertility due to its profuse growth, coppice nature, rapid decomposition rate, and higher nutrient contents. *Gliricidia*-maize/legume intercropping systems sequester more carbon in the soil via continuous application of tree prunings and root turnover. *Gliricidia sepium* can also replenish soil fertility through biological nitrogen fixation and enhance recycling of nutrients in the soil through incorporation of nitrogen-rich leaves as green manure (refer Figure 8.3).

v   *Resilience:* Interplanting of *Gliricidia* and incorporation of its leaf biomass enhance resilience of farming systems to climate change. *Gliricidia sepium* is a drought-resistant tree as it sheds most of its leaves during the dry season, thus reducing water loss at the time of transpiration. When properly incorporating the leaf biomass into the soils, *G. sepium* increases the organic matter content of soils, improves soil aeration, reduces soil temperature, reduces soil erosion, and contributes to weeds control (Akinnifesi et al., 2010). Thus, integrating *G. sepium* in the AC systems will build up resilience to climate change adaptation and mitigation. The farmers in the demo trials implement conservation agriculture (CA) practices that include reduced tillage using animal-drawn tillage implements called ripper, and retain crop residues to cover the soils in ripper lines. These CA practices reduce soil erosion and improve moisture content by avoiding water stress during dry periods (Thierfelder and Wall, 2009). However, there are challenges that hinder widespread uptake of the *Gliricidia* agroforestry technology by small holder farmers. This includes land shortage, insecure land tenure system, lack of tree seeds, and knowledge-intensive nature of the *Gliricidia* agroforestry technology.

Addressing the socio-economic dimension of agroecology

vi   *Co-creation and sharing of knowledge*: Farmers in the study areas used to collect the leaves of *Gliricidia* and apply it as mulch (by spreading the leaves/biomass on the surface of soils) to fertilize their soils. Incorporation of the leaves into the soil was not practised in the study areas due to lack of knowledge. The farmer-led demo trials were used to showcase the benefits of incorporation of leaf biomass into the soils, e.g., in terms of increasing plant-available N, organic matter in the soils, carbon storage (see Figure 8.4), increasing crop yields (refer Table 8.4), and enhancing co-creation of knowledge and resilience to climate change. In this regard, a range of multimedia platforms (e.g., radio broadcast, newspaper, better-life booklets, and video documentaries) were used to increase awareness and disseminate information about the advantages of on-farm *Gliricidia* tree plantings, leaf manure incorporation, and general farm management. These platforms have reached out others like neighbouring farmers, traditional leaders, district officials, and other stakeholders who are not involved in the project. It was possible to reach out to 230,000 small-scale farmers (about 50% of them women) who are currently practising agroforestry in the eastern districts of Zambia where the study was undertaken. Social learning, and integrating scientific and local knowledge were important for increased adoption of AE practices and the development of *Gliricidia* agroforestry systems in the eastern Zambia.

vii   *Human/social values (including gender integration, labour cost)*: Women farmers in the study areas are actively participating in a range of activities such as raising/planting of the seedlings, incorporation of leaf biomass (see Figure 8.3), and participation in leadership at the community level. Agroforestry with *Gliricidia* intervention can empower rural women and smallholders with additional products that generate income. Access to seedlings and water will promote the adoption of agroforestry. It appears that additional farm labour is needed to plant the seedlings, to implement prunings of the coppice, and to incorporate the leaf biomass into the soils. The costs of seedlings and their availability, opportunity costs, and low capacity of women farmers to carry out tree plantings might pose limitations for increased adoption of the AE practices. Although the total cost of *Gliricidia* agroforestry interventions (CA practices, farm inputs inclusive of labour) is challenging in the initial year, the cost is negligible in the subsequent years. It provides multiple benefits in terms of ecological and socio-economic aspects (refer Table 8.2). Once farmers observe the benefits of *Gliricidia* agroforestry, they will be motivated to adopt the technology and build up resilience to climate change.

viii *Policy/governance (measures for increased AE adoption)*: AC systems with agroforestry tree such as *Gliricidia* is one alternative intervention for farmers in the study area to increase AE adoption. However, AE transition requires farmer motivation and capacity (Schoonhiven and Runhaar, 2018). A collective effort is needed between state and non-state agencies/actors to increase the scale of AE adoption. These actors need to create enabling environments through provision of incentives, credit facilities that provide access to quality seeds, and market opportunities such as carbon credits sales. These are channelized through carbon offset scheme (by the government) where communities are then paid for their conservation efforts related to AE practices. In this regard, the Community Markets for Conservation (COMACO: a project partner) is assisting farming communities/cooperatives in the case study districts of eastern Zambia in collaboration with the local government. COMACO is a social enterprise that supports small-scale farmers in Zambia by promoting the adoption of AE practices such as conservation agriculture, AC with agroforestry, and other income-generating activities (e.g., honey production). This shows that interventions by such non-governmental organizations are necessary to promote agroecology approaches.

ix *Culture/food traditions (traditional foods, nutrition quality)*: The small-scale farmers in the study areas are facing food and nutrition insecurity due to a range of ecological and socio-economic factors. Crops produced under AC systems (legumes) with *Gliricidia* are organic as chemical inputs are not added to the soil. Such systems improved the soil health, crop health, and food quality as shown in Tables 8.5–8.7 and qualify for better market opportunities. In contrast, maize-based monocropping systems that rely on external inputs have resulted in poor soil health, lower yield, and poor nutritional quality. Thus, AC with *Gliricidia* plays a vital role in re-balancing traditional and modern food habits by promoting healthy food production and consumption, while ensuring the right to adequate food.

x *Circular solidarity economy (including value chain improvements)*: Social and institutional innovations play a key role in increasing AE production and consumption. One such example is the role played by COMACO in the case study area. COMACO connects producers and consumers, increases the value addition of farmer produce, and opens new markets. The innovative markets respond to consumers' growing demand for healthier diets while encouraging AE production. This approach makes food value chains shorter and more resource efficient. It also reduces food production losses or wastage by enhancing FNS while reducing pressure on natural resources.

Conclusion

This chapter reviewed literature related to the key principles/elements of agroecology (AE) and elaborated their implications to science, practice, and policy. One of the main barriers to adopting AE is the lack of evidence on the interactive effects of the practices on AE elements. The case study (i.e., *Gliricidia* agroforestry project in Zambia) has implemented a range of AE practices and approaches that include intercropping, leaf manure incorporation, residue mulching, and value addition on the AE farming products. The results demonstrated the synergistic effects on adaptation and mitigation to climate change. More specifically, the farmer-led demonstration trials on AC systems with *Gliricidia* agroforestry showed positive impacts on the ecological/environmental and socio-economic dimensions of AE elements and principles:

i   *Ecological dimension of AE elements/principles*: Soil health and crop nutrition were improved by incorporating bio-degradable leaf biomass of *Gliricidia sepium* in the AC systems. Combining maize-legume-agroforestry conservation practices with *Gliricidia* provided multiple benefits and reduced risks to smallholders. The use of AC practice with *Gliricidia* increased the production of nutritious food crops (such as groundnuts and soybeans) and has improved the quality of the crops. It enhanced the overall food and nutrition security and resilience to climate change adaptation and mitigation, as evident from the data.

ii *Socio-economic dimensions of AE elements/principles*: Farmers implemented conservation agriculture practices (reduced tillage using rippers, residue mulching, and crop rotations), as an adaptation strategy to mitigate the effects of erratic rainfall. The *Gliricidia* leaf biomass incorporation into the soils has provided an alternative for small-scale farmers to apply a low-cost organic fertilizer into their soils. The introduction of AC systems with *Gliricidia* agroforestry in the eastern province of Zambia has prompted the adoption of AE farming practices, despite additional labour costs required in the initial year of the tree establishment due to the benefits it generated. In general, the AC systems with *Gliricidia* agroforestry practices proved to be effective on the key element of AE. However, good AE practices that could minimize trade-offs in crop-tree-animal interactions in vulnerable farming systems in different agroecological settings are recommended for further investigation.

Acknowledgement

The authors would like to thank the farmers and field enumerators of the Gliricidia-Zambia project and the NORAD/Ministry of Foreign Affairs, Norway for funding the project.

Note

1 https://www.nibio.no/en/projects/gliricidia?locationfilter=true

References

AGRA (Alliance for a Green Revolution in Africa) (2016) 'Africa Agriculture Status Report 2016: Progress Towards Agricultural Transformation in Africa'. Available at: https://agra.org/aasr2016/public/assr.pdf

Akinnifesi, F.K., Ajayi, O.C., Sileshi, G., Chirwa, P.W. and Jonas, C. (2010) 'Fertilizer Trees for Sustainable Food Security in the Maize-based Production Systems of East and Southern Africa: A Review', *Agronomy for Sustainable Development* , vol. 30, pp. 615–629.

Alamu, E.O., Gondwe, T., Akinwale, G., Suzuki, K., Chisonga, C., Chigeza, G. and Busie, M.D. (2019) 'Impact of Soil Fertility Management Practices on the Nutritional Quality of Soybean (Glycine max (l.) Merr.) Varieties grown in Eastern Zambia', *Cogent Food & Agriculture*, vol. 5, pp. 10–13.

Altieri, M.A. and Nicholls, C.I. (2018) 'Urban Agroecology: Designing Biodiverse, Productive and Resilient City Farms', *AgroSur*, vol. 46, pp. 49–60.

Anderson, C.R., Pimbert, M.P., Chappell, M.J., Brem-Wilson, J., Claeys, P., Kiss, C., Maughan, C., Milgroom, J., McAllister, G., Moeller, N. and Singh, J. (2020) 'Agroecology Now—Connecting the Dots to Enable Agroecology Transformations', *Agroecology and Sustainable Food Systems*, vol. 44 (5), pp. 561–565.

Ayala-Orozco, B., Rosell, J.A., Merçon, J., Bueno, I., Alatorre-Frenk, G., Langle-Flores, A. and Lobato, A. (2018) 'Challenges and Strategies in Place-Based Multi-Stakeholder Collaboration for Sustainability: Learning from Experiences in the Global South', *Sustainability*, vol. 10, p. 3217.

Asibuo, J.Y., Akromah, R., Safo-Kantanka, O., Adu-Dapaah, H.K., Ohemeng-Dapaah, S. and Agyeman, A. (2008) 'Chemical Composition of Groundnut, Arachis hypogaea (L) landraces', *African Journal of Biotechnology*, vol. 7 (13), pp. 2203–2208.

Atasie, V.N., Akinhanmi, T.F. and Ojiodu, C.C. (2009) 'Proximate Analysis and Physico-Chemical Properties of Groundnut (Arachis hypogaea L.)', *Pakistan Journal of Nutrition*, vol. 8(2), pp. 194–197.

Bernard, B. and Lux, A. (2017) 'How to Feed the World Sustainably: An Overview of the Discourse on Agroecology and Sustainable Intensification', *Reg Environ Change*, vol. 17, pp. 1279–1290.

Bezner Kerr, R., Young, S.L., Young, C., Santoso, M.V., Magalasi, M., Entz, M., Lupafya, E., Dakishoni, L., Morrone, V., Wolfe, D. and Snapp, S.S. (2019) 'Farming for Change: Developing a Participatory Curriculum on Agroecology, Nutrition, Climate Change and Social Equity in Malawi and Tanzania', *Agric Hum Values*, vol. 36 (3), pp. 549–566.

Brady, N.C. and Weil, R.R (2002) *The Nature and Properties of Soils*, 13th Edition. Upper Saddle River, NJ: Prentice Hall; 960 p; ISBN: 13-016763-0.

CNS-FAO (2021) 'Pathways to advance agroecology. Overcoming challenges and contributing to sustainable food systems transformation. Swiss National FAO Committee (CNS-FAO), March 2021'. Available at: https://www.oneplanetnetwork.org/sites/default/files/cns-fao_working-document_advancing_agroecology_march_2021.pdf

Deaconu, A., Berti, P.R., Cole, D.C., Mercille, G. and Batal, M. (2021) 'Agroecology and Nutritional Health: A Comparison of Agro-ecological Farmers and Their Neighbours in the Ecuadorian Highlands', *Food Policy* vol. 101, pp. 1–14.

Etiosa, O.R., Chika, N.B. and Benedicta, A. (2017) 'Mineral and Proximate Composition of Soya Bean', *Asian Journal of Physical and Chemical Sciences*, vol. 4 (3), 1–6.

EC (European Commission) (2021) Soil Deal mission implementation plan, section 8B. Available at: https://ec.europa.eu/info/publications/implementation-plans-eu-missions_en

EC (European Commission) (2020) Available at: https://ec.europa.eu/info/news/46-new-projects-start-their-research-agroecology-and-ocean-observation-2020-sep-28_en

FAO (Food and Agriculture Organization) (2018) *The 10 Key Elements of Agroecology – Guiding the Transition to Sustainable Food and Agricultural Systems*. Rome: FAO. Available at http://www.fao.org/3/i9037en/I9037EN.pdf

FAO (Food and Agriculture Organization) (2019) *TAPE Tool for Agroecology Performance Evaluation 2019 – Process of Development and Guidelines for Application, Test Version*. Rome: FAO. Available at: https://www.fao.org/documents/card/en/c/ca7407en/

FAO (Food and Agriculture Organization) (2020) 'FAOSTAT'. Available at: http://www.fao.org/faostat/en/#home

Goudiaby, A.O.K, Diedhiou, S., Diatta, Y., Badiane, A., Diouf, P., Fall, S., Diallo, M.D. and Ndoye, I. (2020) 'Soil Properties and Groundnut (*Arachis hypogea L.*) Responses to Intercropping with Eucalyptus camaldulensis Dehn and Amendment with Its Biochar', *Journal of Materials and Environmental Sciences*, vol. 11, pp. 220–229.

HLPE (High Level Panel of Experts) (2019) 'Agro-ecological and Other Innovative Approaches for Sustainable Agriculture and Food Systems That Enhance Food Security and Nutrition', High Level Panel of Experts on Food Security and Nutrition of the Committee on World Food Security, Rome. Available at: http://www.fao.org/fileadmin/user_upload/hlpe/hlpe_documents/

Horneck, D.A., Sullivan, D.M., Owen, J.S. and Hart, J.M. (2011) 'Soil Test Interpretation Guide', Oregon State University Extension Service, EC 1478, USA. Available at: https://ir.library.oregonstate.edu/concern/administrative_report_or_publications/2b88qc45x

IAASTD (International Assessment of Agricultural Knowledge, Science and Technology for Development) (2009) 'Agriculture at a Crossroads: Global Report', Island Press, Washington, DC.

ILO (International Labor Organization) (2018) 'Decent Work and the Sustainable Development Goals: A Guidebook on SDG Labour Market Indicators', Department of Statistics, Geneva, ISBN 978-92-2-132117-0.

IPCC (Intergovernmental Panel on Climate Change) (2019) 'Climate Change and Land. IPCC Special Report on Climate Change, Desertification, Land Degradation, Sustainable Land Management, Food Security, and Greenhouse Gas Fluxes in Terrestrial Ecosystems'. Available at: https://www.ipcc.ch/srccl/

IPES-Food (2016) 'From Uniformity to Diversity: A Paradigm Shift From Industrial Agriculture to Diversified Agro-Ecological Systems', *International Panel of Experts on Sustainable Food Systems*. Available at: https://ipes-food.org/reports/

Lampkin, N.H., Pearce, B.D., Leake, A.R., Creissen, H., Gerrard, C.L., Girling, R., Lloyd, S., Padel, S., Smith, J., Smith, L.G., Vieweger, A. and Wolfe, M.S. (2015) 'The Role of Agroecology in Sustainable Intensification', Report for the Land Use Policy Group, Organic Research Centre, Elm Farm and Game & Wildlife Conservation Trust. Available at: http://publications.naturalengland.org.uk/publication/6746975937495040

Leippert, F., Darmaun, M., Bernoux, M. and Mpheshea, M. (2020) 'The Potential of Agroecology to Build Climate-resilient Livelihoods and Food Systems', Rome. FAO and Biovision. Available at: https://doi.org/10.4060/cb0438en

Makumba, W., Janssen, B., Oenema, O., Akinnifesi, F.K., Mweta, D. and Kwesiga F (2007) 'The Long-term Effects of a Gliricidia-maize Intercropping System in Southern Malawi on Gliricidia and Maize Yields and Soil Properties', *Agriculture, Ecosystems & Environment*, vol. 116, 85–92.

Moebius-Clune, B.N., Moebius-Clune, D.J., Gugino, B.K., Idowu, O.J., Schindelbeck, R.R., Ristow, A.J., van Es, H.M., Thies, J.E., Shayler, H.A., McBride, M.B., Kurtz, K.S.M., Wolfe, D.W. and Abawi, G.S. (2016) *Comprehensive Assessment of Soil Health – The Cornell Framework*, Edition 3.2. New York: Cornell University.

Mockshell, J. and Kamanda, J. (2018) 'Beyond the Agro-ecological and Sustainable Agricultural Intensification Debate: Is Blended Sustainability the Way Forward?', *International Journal of Agricultural Sustainability*. Available at: https://doi.org/10.1080/14735903.2018.1448047

Newton, P., Civita, N., Frankel-Goldwater, L., Bartel, K. and Johns, C. (2020) 'What Is Regenerative Agriculture? A Review of Scholar and Practitioner Definitions Based on Processes and Outcomes', *Frontiers in Sustainable Food Systems*, vol. 4, p. 577723.

Niggli, U., Sonnevelt, M. and Kummer, S. (2021) 'Pathways to Advance Agroecology for a Successful Transformation to Sustainable Food Systems', *Food Systems Summit Brief prepared by Research Partners of the Scientific Group*. Available at: http://doi.org/10.48565/

Nguyen, H. (2018) 'Sustainable Food Systems Concept and Framework', Food and Agriculture Organization of the United Nations: Rome, Italy. Available at: https://www.fao.org/publications/card/en/c/CA2079EN/

Ogunyemi, A.M., Otegbayo, B.O. and Fagbenro, J.A. (2018) 'Effects of NPK and Biochar Fertilized Soil on the Proximate Composition and Mineral Evaluation of Maize Flour', *Food Science & Nutrition*, vol. 6, pp. 2308–2313.

Parmentier, S. (2014) 'Scaling-up Agro-ecological Approaches: What, Why and How?' Oxfam Solidarité.

Pretty, J.N., Toulmin, C. and Williams, S. (2011) 'Sustainable Intensification in African Agriculture', *International Journal of Agricultural Sustainability*, vol. 9 (1), pp. 5–24.

Schader, C., Grenz, J., Meier, M. and Stolze, M. (2014) 'Scope and Precision of Sustainability Assessment Approaches to Food Systems', *Ecology and Society*, vol. 19 (3), p. 42.

Schoonhiven, Y. and Runhaar, H. (2018)' Conditions for the Adoption of Agro-ecological Farming Practices: A Holistic Framework Illustrated With the Case of Almond Farming in Andalusia', *International Journal of Agricultural Sustainability*, vol. 16 (3), pp. 1–13.

Sida, T.S., Baudron, F., Hadgu, K., Derero, A. and Giller, K.E. (2018) 'Crop vs. Tree: Can Agronomic Management Reduce Trade-offs in Tree-crop Interactions?' *Agriculture, Ecosystems & Environment*, vol. 260, pp. 36–46.

Struik, P.C. and Kuyper, T.W. (2017) 'Sustainable Intensification in Agriculture: The Richer Shade of Green. A Review', *Agronomy for Sustainable Development*, vol. 37, p. 39.

Thierfelder, C. and Wall, P.C. (2009) 'Effects of Conservation Agriculture Techniques on Infiltration and Soil Water Content in Zambia and Zimbabwe', *Soil & Tillage Research Journal*, vol. 105, pp. 217–227.

von Braun, J., Afsana, K., Fresco, L.O. and Hassan, M. (eds.) (2021) 'Science and Innovation for Food Systems Transformation and Summit Actions', Papers by the Scientific Group and its partners in support of the UN Food Systems Summit. Available at: https://sc-fss2021.org

Wezel, A. and Silva, E. (2017) 'Agroecology and Agro-ecological Cropping Practices', pp. 19–51: In: *Agroecological Practices for Sustainable Agriculture: Principles, Applications, and Making the Transition*, World Scientific, New Jersey, USA, 978-1-78634-305-5.

9 Transitioning toward climate-neutral and resilient smallholder farming systems

An institutional perspective

Giacomo Branca, Luca Cacchiarelli, Udaya Sekhar Nagothu, and Chiara Perelli

Introduction

The global food production system faces many challenges, including increasing food demand due to a growing population and climate change, which is expected to affect food production and stress the natural resource base upon which agriculture depends (IPCC, 2014). This is particularly true in sub-Saharan Africa, where a fast-growing population, food insecurity, environmental degradation, resource depletion, and increasing smallholder vulnerability to climate change is making it difficult to scientists and policy makers to address the problems (Li et al., 2019). For African smallholders, it is even more important to adopt climate-resilient agriculture in order to make a sustainable transition toward climate-neutral and resilient farming systems (CNRFS). However, the adoption and diffusion of climate-smart technologies have been slow (Branca and Perelli, 2020). The underdeveloped rural financial options, inadequate research and extension services, insufficient market infrastructure, and lack of policy support often contribute to the slow diffusion of innovation in the agriculture sector.

Value chains (VCs) represent one of the few options for small producers to access larger markets and innovative technologies (World Bank, 2007). However, the private sector does not see the smallholder segment as a potential market source for its products and services and vice versa. Indeed, most smallholders in developing countries face bottlenecks in accessing markets and in capturing the value addition, which is often exploited by intermediaries along the VC. Unlocking the complexity in VC pathways, strengthening linkages among the different actors of the VCs, and supporting the development of innovative business models for small producers can contribute to overcome such barriers to market entry. This is particularly relevant for the smallholder adoption of CNRFS-related innovations.

The objective of the chapter is to provide an institutional perspective about innovations for a transition toward CNRFS, with a focus on VCs. In highlighting the role played by stakeholders in the dynamics and partnerships for the diffusion of climate-resilient innovative technologies, we focus on the how and who should

DOI: 10.4324/9781003273172-9

be engaged, and what are the benefits and challenges of such engagement. In this context, the case of dairy VCs in two Eastern Africa countries (Kenya and Rwanda) will be discussed, with focus on the socio-economic barriers faced by smallholders. Adoption of technological innovations is dependent on the proper institutional and policy support. The recommendations from the chapter can help in developing frameworks for upscaling adoption of CNRFS. Right policy and institutional settings are necessary to overcome barriers to innovation adoption, and to foster coordination.

The chapter is organized as follows: the next section presents the conceptual framework. The case studies are described in section "Case studies", followed by the results in section "Discussion". Toward the end, conclusions are presented.

Conceptual framework

Multiple institutional factors can prevent primary producers from adopting innovative technologies and, in turn, exploiting market opportunities and the business environment (Poulton et al., 2006; Markelova et al., 2009; Nagothu, 2015, 2018). They include (i) households' socio-economic characteristics, including their assets, education, gender, and property rights; (ii) limitations in infrastructure and input markets, for instance, credit, seed, or fertilizer; and (iii) insecure access to information services.

Smallholder farmers' decision to adopt agricultural innovations requires a good combination of the institutions and policies, which can help to overcome barriers and limiting factors. From an institutional perspective, different models of VC integration are possible (Montefrio and Dressler, 2019), ranging from informal agreements to more complex and formalized relations such as out-grower schemes (Branca et al., 2016). The VC partnerships are increasingly becoming useful pathways to tackle these limitations, evidenced in the active promotion of multi-stakeholder groups represented by the different VC actors – for instance, producers, farmer organizations, input and service providers, private sector, research institutions, government agencies and non-governmental organizations (NGOs), small and medium enterprises (SMEs) that operate at different levels. The synergy derived from the partnerships can overcome barriers in the adoption (Kolk et al., 2008). Partnerships should be based on interactive learning, empowerment, and collaborative governance that enables stakeholders with interconnected problems and ambitions, but with different interests, to be collectively innovative and resilient when faced with the emerging risks, crises, and opportunities of a complex and changing environment (Woodhill and van Vugt, 2011). By addressing the institutional business environment, partnerships can play a pivotal role in enhancing the chances for producers to be viable suppliers of VCs being a combination of organizational activity functional to production and marketing (Wijk et al., 2010). Partnerships can be vehicles for the diffusion of agricultural innovations (Hermans et al., 2017).

Successful cases of innovation adoption invariably demonstrate a range of partnerships and network-like arrangements that connect knowledge users, knowledge producers, and others involved in the market, policy, and civil society arenas (Hall, 2012). In this context, public extension services can play a brokerage role, beyond their traditional role of linking technology and farmers, networking with relevant VC actors, and can help to negotiate changes in the policy environment and investment arrangements. Several factors and processes enable or hinder interactions, both within and external to multi-actor co-innovation partnerships (Cronin et al., 2021). Factors that enable partnerships to achieve their own goals are based on the inclusion of partners linked with already existing networks that can facilitate internal collaboration and couple with external environment including policy and market conditions.

Smallholder farmers need to be genuinely engaged with the VC actors so that they benefit from the added value for their products (AFI, 2017). The success of a particular product in VC development will depend on smallholder stewardship of the program and their involvement early in the VC development process (CGI, 2016). On their own, small farmers who constitute a majority are disadvantaged when it comes to accessing markets, credit, and agricultural resources. This is one of the reasons for poor adoption of innovations on small farms. In response, countries such as India have initiated Farmer Producer Organizations (FPOs) to enable farmers work collectively to reduce costs, improve market access, drive higher agricultural productivity, enhance food security and livelihood development (Verderosa, 2021). The FPOs provide a good platform for strengthening smallholder stewardship in the VC development.

From the policy point of view, a stable political environment with adequate legislative measures can favor innovation adoption and encourage rural revitalization (Kosec and Resnick, 2019; Branca et al., 2022). A wide variety of options exist to create a policy environment conducive to innovation adoption (Lybbert and Sumner, 2012), ranging from legislative and regulatory instruments to direct investments, property right allocations, and economic incentives or subsidies.

The adoption of CNRFS will succeed when there are stable and assuring markets for the farmer's produce also providing adequate opportunity to farmers to earn higher incomes. The extent of adoption will also depend on social and environmental context, whether farmers are educated and used to new tools and knowledge, age and gender (Nagothu, 2018). It is important to consider whether the knowledge transfer takes into proper consideration factors such as gender with differentiated needs. A transformative change of smallholders toward CNRFS is required to cope with climate change and ensure food and nutrition security. Climate-resilient innovative farming practices could include (i) improved agronomic practices and effective crop management, (ii) tillage and residue management, and (iii) efficient water management. A combination of improved agronomic technologies and practices can be used to cope with the more unpredictable conditions and the resulting impacts caused by climate change. Examples of such technology packages comprise use of improved crop varieties (e.g., heat and pest tolerant), implementation of crop rotation or intercropping (e.g., cereal-legume), planting cover crops, and avoiding bare fallow (Scialabba et al., 2010). Tillage cropping systems focus on minimum soil disturbance in conjunction with the retention of crop residues on the soil surface (mulching) to enhance water infiltration, prevent runoff, and protect the soil from erosion and crusting by rainfall (Scopel et al., 2004). Proper water management can help capture more rain, making more water available to crops, and using water more efficiently (Rockstrom and Barron, 2007; Vohland and Barry, 2009; Branca et al., 2013), e.g., through planting pits and tied ridge systems which increase infiltration, reduce erosion and the loss of water and soil from arable land (Wiyo et al., 2000). Such conceptual links are shown in Figure 9.1 and discussed below.

Households' socio-economic characteristics. Socio-economic characteristics of smallholder producers are highly heterogeneous (de Oca Munguia and Llewellyn, 2020). Their capacities can be different in terms of education and knowledge intake. According to Huffman (2020), innovation adoption is facilitated by enhanced knowledge and access to formal education which may improve human capital and management capacity. Besides, assuring physical assets' property rights (e.g., land tenure) can help farmers obtain long-term benefits from current investments, thereby increasing the likelihood of adoption (Kassie et al., 2015; Mwangi and Kariuki, 2015; Branca and Perelli, 2020). In this context, social capital (e.g., inclusion in a social network) facilitates innovation adoption, especially on smallholder farms (Husen et al., 2017). Social capital cannot ignore the importance of women and their contribution to agriculture. However, agricultural research and extension has been traditionally biased toward men and there has not been an

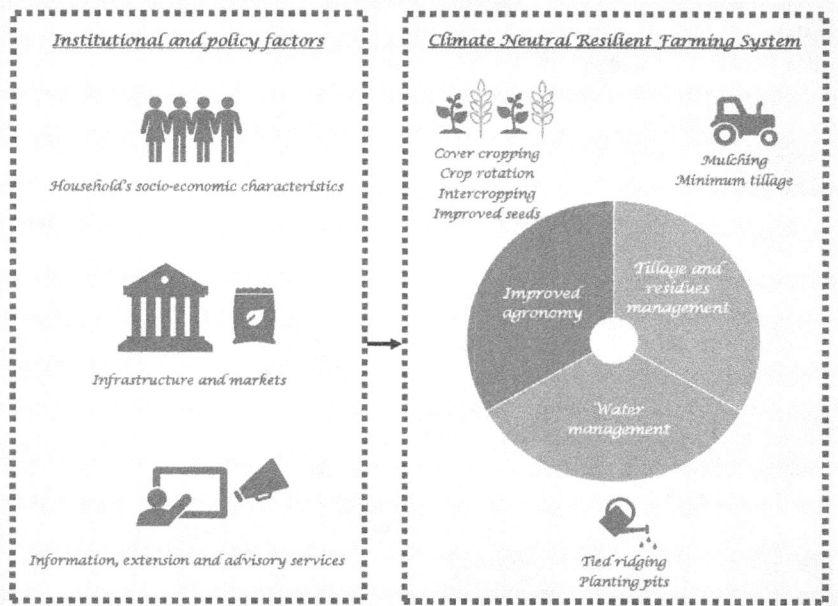

Figure 9.1 Institutional and policy factors affecting smallholders' adoption of CNRFS-related innovations: a conceptual framework.

adequate focus on women (Nagothu, 2015). This is a major challenge in societies where the gender divide is large, and women are not allowed to make decisions on par with men. CNRFS innovations that ignore the specific needs of women can be a major setback to ensure the success of adoption. This can become a major concern where out-migration of men is happening due to economic, climatic, and other reasons, which leaves the burden of agriculture on women. This is the case in several developing countries, including parts of Africa, where men and youth are migrating to cities, as agriculture is becoming risky and is no longer profitable.

Infrastructure and markets. Poor infrastructure and marketing services, costly inputs and transportation, limited access to output markets, and inadequateness of post-harvest facilities (e.g., storage and agro-processing options) represent critical barriers to vertical coordination, preventing smallholders' market access and value addition (Barrett, 2008). This can influence producers' capacity and propensity to make investments in technology innovations, and to determine the appropriate innovation strategy (Mutenje et al., 2016). Most smallholders in Africa are not linked to markets due to various reasons including their remoteness in location, low farm-gate prices, and lack of organization (Wiggins and Keats, 2013). Often a catalyst is necessary to establish linkages between farmers and markets or organize them into groups or collectives. In any case, functioning and accessible markets, particularly for agricultural commodities, are vital for agricultural growth to realize its potential as a powerful driver of rural poverty reduction (Kürschner et al., 2016). Since farming is a risky business, planning and development of VCs should consider all possible risks, including market and political, to ensure that adequate mitigation measures are in place.

Information, extension, and advisory services. Access to information and knowledge about agricultural innovations is another limiting factor of technology adoption (Cafer and Rikoon, 2018). Public extension and advisory services (EAS) often offer low-quality services but the increased private sector involvement in public agricultural extensions (e.g., through public-private partnerships) may also leave resource-poor farmers underserved (Birner et al., 2009; Branca et al., 2022). Advisory services should be designed to facilitate smallholder households' access (Norton and Alwang, 2020) and to link technology adoption to enhancement of market opportunities (Haug et al., 2021). Digital extension tools are being increasingly used these days to bridge the knowledge, gender, and digital divides and empower the rural community by fostering inclusive development and participatory communication (Raj and Nagothu, 2016). An innovative example of digital extension platform is the village knowledge center that can facilitate timely dissemination of knowledge through multiple communication tools.

Case studies

This section presents the results of two case studies in Eastern Africa (Kenya and Rwanda).[1] The case studies explore the socio-economic, physical, and agro-ecological factors that influence on-farm adoption of innovative climate-smart agricultural practices and the related adoption barriers, with a focus on the VC and the relevant institutional and policy perspectives. This chapter focuses on the introduction of the *Brachiaria* grass forage to improve the livestock dairy value chain and the factors that influenced its success in adoption. The forage is an innovation in the case study areas and has contributed to increased climate resilience of the current dairy production systems, fostering their transition toward CNRFS.

Brachiaria grass is a perennial tropical forage with high productivity in terms of palatable and nutritious biomass, tolerates abiotic and biotic stresses, improves soil fertility, produces more nutritious animal feed, and increases overall livestock productivity (Mutimura and Ghimire, 2020). In addition to improvements in livestock productivity in terms of milk production, it is known to contribute significantly toward ecological restoration of degraded lands and soil erosion control (Ghimire et al., 2015). With its appositive traits, it can be one of the promising climate-neutral resilient forage and a good component that can strengthen adaptation and mitigation in crop-livestock integrated systems. In Rwanda, the *Brachiaria* grass has proved to improve the resilience of mixed crop-livestock systems and a buffer against frequent crop failures due to extreme weather events and climate change (Mutimura and Ghimire, 2020). It has large root systems, sequesters carbon into soils, resistant to droughts, performs well in low fertility soils, and provides several environmental benefits in the form of ecosystem services (Djikeng et al., 2014; Njarui et al., 2016, 2020). The fodder grass has a positive impact not only on milk production but also on crop yields (in crop rotation systems) due to the benefits it has on soil fertility. Overall, it generates significant ecological, nutritional, and socio-economic benefits (Table 9.1).

The introduction of *Brachiaria* grass in the farming systems of the case study areas has been achieved due to the promotion of a participative value chain governance approach supported by multi-actor platforms (MAP) established in the two cases, i.e., a partnership aimed at linking farmers' organizations, scientific community, public and private sector, non-governmental organizations (NGOs), and SMEs operating within the same product chain. The MAP members have been involved in different activities, including validation, extension, providing feedback, and upscaling of innovative *Brachiaria*-based dairy production systems. Experience has shown that the MAP members played an important role in strengthening science-policy linkage.

Table 9.1 Ecological, nutritional, and socio-economic benefits provided by *Brachiaria* forage

Ecological benefits	Nutritional benefits	Socio-economic benefits
• Increased livestock productivity • Reduced GHG emissions • Greater climate resilience	• Increased dairy cattle and beef productivity • Improved household nutrition and health	• Increased household income and improved livelihood • Adaptation through income diversification

Adapted from Ghimire et al. (2015).

Evidence shown here is supported by primary data collected through sample household surveys and focus group discussions, MAP meetings, and key inform-ant interviews conducted in the project sites from 2018 to 2020 through a H2020 project "InnovAfrica" funded by the European Commission (www.innovafrica. eu). For each case study, we provide a list of actors operating at various stages of the VC; a description of smallholder farmers' characteristics, which are expected to have an impact on the innovation adoption process (and descriptive statistics resulting from the household survey); a catalog of the policies and institutions in place; and a narrative about the potential strategies to overcome adoption barriers along the VC, developed through the *Theory of Change* methodology through MAP meetings.

The case study in Kenya

The study was carried out in the Kangundo subcounty, which is situated in the eastern midlands. Most smallholders rely on agriculture income from the grow-ing of maize and grain legumes and from livestock production. The introduction of *Brachiaria* grass forage into the dairy cattle production system was expected to generate benefits on biomass production and livestock productivity and, indi-rectly, on smallholders' livelihood and food security. The dairy VC is described below. It was centered on small producers and comprises input providers, traders, processors, and retailers.

The value chain structure

Input provision: Various entities supply inputs and services to farmers and dairy cattle herders. Agro-dealers sell seeds, fertilizers, and pesticides. The most pre-ferred forage seed attributes are pasture quality, suitability for area/local climatic conditions, and durability in terms of harvesting period. Sales arrangements in-cluded cash, credit, discounted for bulk, and discounted for preferred customers. Fertilizers were subsidized under the National Accelerated Agricultural Inputs Access Program (NAAIAP). Agro-dealers were appointed as distribution agents for subsidized inputs by the government. To sell seeds, a license is required, in ad-dition to a business permit provided by the county government. Agro-vet compa-nies and aggregators (cooperatives) supply feeds, supplements, drugs, and artificial insemination (AI) services. They usually make a one-on-one contact with farmers as well as site visits (in collaboration with extension agents). The government re-quires that all agro-vets have an attendant trained in veterinary (e.g., a para-vets or veterinaries registered with the Veterinary Board).

Output production: Small-scale dairy farmers account for about 80% of produc-ers. On average, they own one milking cow per family with a calf per cow per year. Daily milk production was about 5–7 liters per day per cow. Farmland size is 1,500 m^2/family, which also includes both animal housing and family house. Land ded-icated to cropland is approximately one acre per farmer household. The remain-ing 20% of farmers are of medium and large scale. Productivity was estimated at

20–30 liters per day from three to four milking cows per household for medium-scale farmers, and at more than 150 liters per day from about ten milking cows for large-scale producers.

Trading: Our study showed that raw milk was sold to aggregators/cooperatives (32%), traders (23%), retailers (4%), and local basic processors (7%). Remaining quantity was directly sold to consumers. Dairy cooperatives allowed farmers to collectively market their produce and access various inputs and services as described below:

a   *Limuro Dairy Cooperative*, with around 8,000 active members. Services provided to members included raw material collection, processing and marketing, subsidized fertilizer provision, extension and technical services (e.g., veterinary, agriculture extension). Service provision to members is based on a credit-system (i.e., costs are charged at the end of each month and deducted directly from milk sales).
b   *Kambusu Cooperative* is the largest cooperative in the area. It collects approximately 3,000 liters per day. Payments to members were made monthly through a bank. Milk was mostly sold outside the area, while the remaining 30% was sold to local retail shops.
c   *Kakuyuni Cooperative* was recently established. Its members are mainly small-scale farmers.

Local traders connected farmers to milk outlets. They mainly comprised milk hawkers who collected milk from farmers and supply to different buyers, including hotels and schools. This marketing channel is preferred by farmers because of prompt payments.

Processing: Processors purchased raw milk directly from individual farmers (e.g., New Kenya Cooperative Creameries) or from farmer cooperatives. The latter option reduces transport and logistics costs. Processing consists of pasteurization and ultra-heating. Milk is then either packed into packets/containers or further processed into yoghurts, butter, cheese, and ghee.

Retailing: Retailers include supermarkets, milk dispensing machines (ATMs), mobile vendors, milk kiosks, and bars. Supermarkets sold a diverse range of dairy products and can operate ATMs which were also operated by individual entrepreneurs. Mobile vendors sold milk to shops, outlets, and small hotels, using private means of transport (motorcycles or bicycles). Milk kiosks or bars sold milk to consumers on behalf of shops or hotels.

Farmers' characteristics

The results of the survey conducted over a sample of 316 households indicated that only 11% of the households in the study area included *Brachiaria* into their farming systems. Table 9.2 reports the socio-economic characteristics of the sample. Most smallholder farmers are male, middle aged. They attended at least primary school. With reference to economic assets, households' average monthly

Table 9.2 Socio-economic and physical characteristics of farmers in Kangundo (n = 316)

Variables	Description	Mean	Std. Dev.	Min	Max
Brachiaria use	Household uses *Brachiaria* (1/0)	0.117	0.322	0	1
Household head characteristics					
HH head gender	Household head male (1/0)	0.759	0.428	0	1
HH head age	Age of household head (years)	57.946	12.935	27	90
HH head education	Household head attended at least primary school (1/0)	0.975	0.157	0	1
Economic assets					
HH total income	Total farm income (USD)	233.724	232.503	4.95	2376
Credit	Access to credit (1/0)	0.320	0.467	0	1
Subsidy fertilizers	Subsidy's access to buy fertilizers (1/0)	0.022	0.147	0	1
Physical assets					
HH total area	Total farm size (ha)	1.815	8.102	0.1	141.7
Local breed	Household own local breed (1/0)	0.427	0.495	0	1
Exotic breed	Household own exotic breed (1/0)	0.310	0.463	0	1
Crossbreed	Household own crossbreed (1/0)	0.472	0.500	0	1
Fertilizers use	Household uses of fertilizers (1/0)	0.665	0.473	0	1
Environmental context					
Semi-arid AEZ	Agro-ecological Zone semi-arid (1/0)	0.981	0.137	0	1
Drought experience	Household experienced drought (1/0)	0.911	0.285	0	1
Flood experience	Household experienced floods (1/0)	0.025	0.157	0	1
Irregular rain experience	Household experienced irregular rains (1/0)	0.873	0.333	0	1
EAS					
Extension provided by government	Access to extension services provided by government	0.206	0.405	0	1
Extension provided by private company	Access to extension services provided by private company	0.044	0.206	0	1
Extension provided by NGO	Access to extension services provided by NGO	0.016	0.125	0	1
Extension provided cooperatives/farmers	Access to extension services provided by cooperatives/farmers	0.098	0.298	0	1
Extension provided by bank/insurance	Access to extension services provided by bank/insurance	0.076	0.265	0	1
Group participation	Participation to groups (1/0)	0.449	0.498	0	1
Household food security					
Food security	Food Consumption Score	85.060	15.747	30	112

Based on own survey data collected.

income amounted to US$233.72. Nearly 32% of farmers had access to credit, while only 2% benefited from input subsidies. Considering the physical assets, the average land available to farmers is less than 2 hectares; most common livestock species are local and crossbreeds; fertilizers were commonly used. With reference to the environmental context, most farmers perceived climate alterations such as droughts (91%) and irregular rains (87%). With regard to social capital, about half of sampled households participated in some form of agricultural groups or associations.

Policies and institutions

Policy context. The main policies implemented in the sector were as follows:

a   Seed and fertilizer subsidies through the "Input policy". They aim to provide inputs to farmers at affordable prices, therefore expanding inputs access to smallholders. However, inadequate funds allocated to this policy and targeting difficulties limited policy effectiveness.
b   Public extension service support through the "Livestock policy". It aims to facilitate demand-driven extension services and increase production efficiency even if farmers lack awareness of its importance. The limiting factors to policy effectiveness were inadequate financial resources allocated to policy, limited capacity of extension workers, and lack of transportation means to reach rural areas.
c   Provision of small irrigation equipment through the "Agriculture irrigation policy". It aims to provide irrigation infrastructure to farmers in arid and semi-arid land. Inadequate financial resources, scarce technologies, and insufficient capacity of technical staff to facilitate implementation limited policy effectiveness.
d   Establishment of appropriate storage facilities through the "Agribusiness policy". It aims to provide storage facilities, make livestock commercially oriented and competitive, and provide capacity building on agribusiness skills. Inaccessibility of appropriate storage facilities (e.g., coolers), limited funds, and insufficient awareness regarding the efficient handling of post-harvest agricultural produce were found to be the main limiting factors.

Extension services: Extension service provision is guided by the National Agriculture Sector Extension Policy (NASEP). It is emphasized that the private sector should play a large role in providing extension services. Despite such a policy, the extension personnel to farmer ratio remained low, the main provider being the public sector. Also, budgetary allocation to extension services has dwindled, and the quality of private extension is questionable. To enhance access to markets, cooperative movements are promoted, but it is not adequate in their current form.

Market and other Institutions: More than three-quarters of sampled farmers had access to the market through traders, cooperatives, and individuals. Low prices and unstable prices were the most important constraints in marketing. Several institutions supported the dairy sector including the Kenya Dairy Board, responsible for ensuring efficient production, marketing, distribution, and supply of milk and dairy products; the Kenya Bureau of Standards (KEBS), with the responsibility of setting and enforcing standards for all products; the Public Health Division of the Ministry of Health, which ensured maintenance of hygiene standards along the chain; and the Kenya Agriculture and Livestock Research Organization, which is incharge of public agricultural research. The combined support from the government agencies plays an important role in strengthening the livestock value chain.

Value chain development strategies

The *Theory of Change* developed was based on inputs from the MAP meetings where members identified the following barriers to the adoption of *Brachiaria*, including lack of information on forage grass, expensive forage seeds due to high production costs and limited seed production capacity, small land size limiting the possibility to develop fodder production, and lack of irrigation opportunities and reliance on fluctuating rainfall patterns. Establishing knowledge platforms to share information might overcome the lack of knowledge about forage grass in general and *Brachiaria* in particular. Use of alternative propagation methods (e.g., splits) and wider involvement of public-private partnerships to multiply seeds are plausible interventions to enhance *Brachiaria* multiplication. Increasing farmers' knowledge about water harvesting techniques and mapping areas indicating suitable locations for irrigating pastures can support forage production expansion.

Smallholders' access to inputs can be increased by expanding subsidy access to a wider range of inputs (seeds, fertilizers, equipment) from a variety of providers, e.g., through the e-voucher digital service delivery. An increased efficient use of inputs might also be achieved through enhancing public-private extension and advisory services and strengthening linkages between research and farmers through innovation sharing. A summary of the Theory of Change exercise is reported in Figure 9.2.

The case study in Kenya

The case study refers to *Nyamagabe* district, situated in the Southern Province, characterized mainly by maize-cattle-based farming systems.

The value chain structure

Input provision: Agro-dealer and agro-vet companies supplied inputs and services. Some agro-dealers also provided technical assistance to farmers, together with public extension agents. Inputs' selling prices are partially set by the Ministry of Agriculture and Animal Resources (MINAGRI) under the subsidy input policy. However, most farmers cannot afford purchasing inputs, even if subsidized, due to income constraints.

Output production: Dairy farmers are mostly (about 90%) small-scale producers with an average of one to two dairy cows per household. Large dairy farmers owned on average six dairy cows. Milk productivity was about 3 liters per cow per day. Production within both small- and large-scale farming was based on integrated crop-livestock systems. Dairy farmers sold milk to dairy cooperatives (30%), local consumers (20%), and local traders (5%), while self-consumption varied between

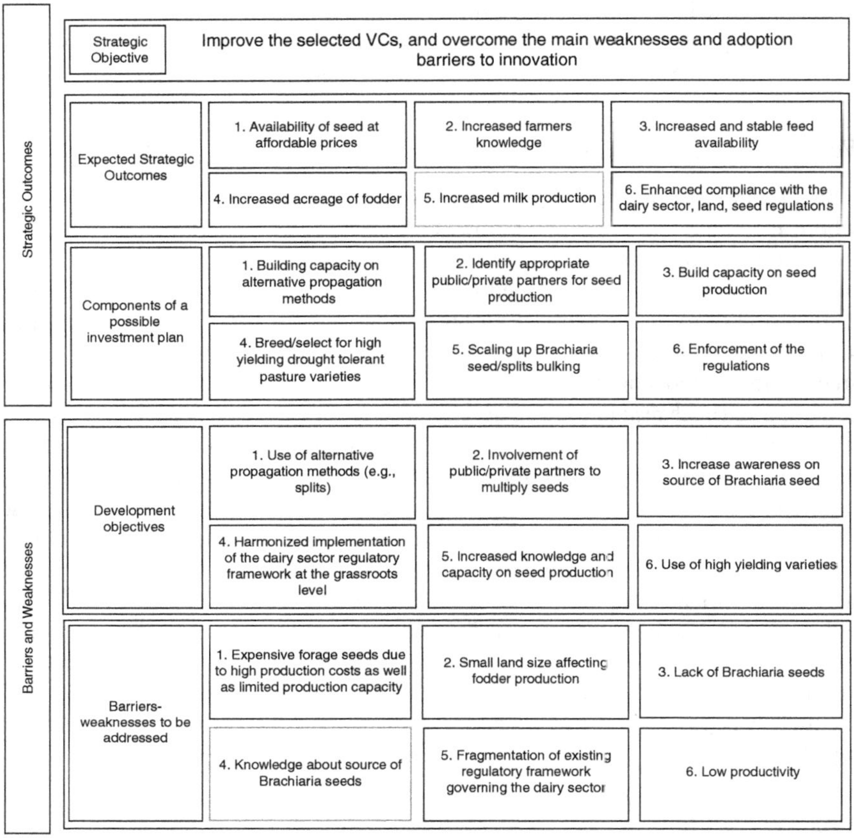

Figure 9.2 Theory of Change: introducing *Brachiaria* into the dairy VC in Kangundo (Kenya).
Source: Authors' own elaboration.

30% and 50%. The minimum price for milk at the farm gate was set by the government and was 200 Rwandan franc (RWF) per liter[2] in 2017. However, on the informal market, the price for local consumers was 160 RWF per liter in the same year. Farmers did not process raw material for the formal market, but they processed fermented milk for self-consumption and for sales to local consumers.

Trading: One dairy cooperative collected the raw product through a milk collection center. With reference to 2017 (when the data collection has been conducted), despite its milk collection capacity of 5,000 liters per day, local dairy farmers supplied only between 500 and 800 liters per day (in the dry and wet season, respectively). The cooperative bought raw milk from local dairy farmers (at a price of 200 FRW per liter) and from local traders (at 220 FRW per liter). The collected and cooled milk was sold at 250 FRW per liter to local traders, restaurants, and single consumers. Local traders in the area operated at two different levels, playing an intermediary role in two stages of the supply chain: (i) they bought milk from farmers at 200 FRW per liter and resold it to the dairy cooperative at 220 FRW per liter; (ii) they bought milk from the dairy cooperative at 250 FRW per liter and resold it to local supermarkets and local restaurants at 300 FRW per liter.

Processing: Packed milk was supplied by national processors, who bought milk from farmers located in other production areas. The largest national company was *Inyange Industries*, which processes and distributes most produced milk in the country. It processed a wide variety of dairy products (packed milk, pasteurized milk, flavored milk, ghee, butter, yoghurt). Products were also exported to Sudan, South Africa, Uganda, Kenya, and Tanzania. Within the domestic market, dairy products were supplied to retailers via independent or own distributors.

Retailing: Independent distributors were registered with the *Inyange Industries*. They bought packed milk at 880 FRW per liter from national processors and sold it to the groceries/supermarket at 930 FRW per liter. Raw milk was bought by distributors at 350 FRW per liter and sold to local groceries at 400 FRW per liter. At the retail level, consumer prices were 1,000–1,200 FRW per liter for packed milk and 430 FRW per liter for raw milk (consumers bring their own containers). Distributors sold packed milk on various markets in the country, whereas individuals mostly sold unpacked cooled milk mainly in urbanized and business center areas.

Farmers' characteristics

The results of the survey conducted over a sample of 308 households indicated that only 4% of the households in the study area included *Brachiaria* into their farming systems. Table 9.3 reports the socio-economic characteristics of the sample. Most smallholder farmers were male, middle-aged, and attended at least primary school. Considering the economic assets, the average monthly income amounted to US$43.9. Almost 40% of sampled farms had access to credit, while only a few farmers benefited from seed and fertilizer subsidies (2% and 3%, respectively). Considering physical assets, the average land parcel size was less than 1 ha; dairy cattle production relied mostly on crossbreeds (79%); fertilizers used was limited (only 13% of sampled farmers). With reference to climate change, most farmers perceived climate alterations, mainly droughts (85%). Approximately 30% of farmers were part of agricultural groups/associations.

Policies and institutions

Policy context: The main policy instruments implemented were as follows:

- A gradual increase in the number of improved dairy cows bred was promoted by the government through the "One cow per poor family" program, whose objectives included fighting malnutrition and poverty through productivity increase and a reduction of pressure caused by grazing on the limited pasture resources.

Table 9.3 Household socio-economic and physical characteristics in Nyamagabe
(308 HHs)

Variables	Description	Mean	Std. Dev.	Min	Max
Brachiaria	Household uses *Brachiaria* (1/0)	0.020	0.057	0	1
Household head characteristics					
HH head gender	Household head male (1/0)	0.714	0.452	0	1
HH head age	Age of household head (years)	52.237	12.418	21	89
HH head education	Household head attended at least primary school (1/0)	1.049	5.617	0	99
Economic assets					
HH total income	Total farm income (USD)	43.975	48.833	0	300
Credit	Access to credit (1/0)	0.377	0.485	0	1
Subsidy seed	Subsidy's access to buy seeds (1/0)	0.019	0.138	0	1
Subsidy fertilizers	Subsidy's access to buy fertilizers (1/0)	0.029	0.169	0	1
Physical assets					
HH total area	Total farm size (ha)	0.693	0.879	0	5.7
Local breed	Household own local breed (1/0)	0.188	0.392	0	1
Exotic breed	Household own exotic breed (1/0)	0.078	0.268	0	1
Crossbreed	Household own crossbreed (1/0)	0.792	0.406	0	1
Fertilizers use	Household uses fertilizers (1/0)	0.133	0.340	0	1
Brachiaria	Household uses *Brachiaria* (1/0)	0.003	0.057	0	1
Environment					
Drought experience	Household experienced drought (1/0)	0.854	0.354	0	1
Floods experience	Household experienced floods (1/0)	0.172	0.378	0	1
Irregular rain experience	Household experienced irregular rains (1/0)	0.169	0.375	0	1
Social assets					
Extension provided by government	Access to extension services provided by government	0.058	0.235	0	1
Extension provided by private company	Access to extension services provided by private company	0.013	0.113	0	1
Extension provided by NGO	Access to extension services provided by NGO	0.003	0.057	0	1
Extension provided by cooperatives/farmers	Access to extension services provided by cooperatives/farmers	0.026	0.159	0	1
Extension provided by bank/insurance	Access to extension services provided by bank/insurance	0.000	0.000	0	1
Group participation	Participation to groups (1/0)	0.315	0.465	0	1
Household food security					
Food security	Food Consumption Score	46.979	19.357	0	100

Based on own survey data collected.

- Subsidies on livestock inputs (AI, forage, seeds) were included in the "Strategic and investment plan to strengthen the animal genetic improvement in Rwanda" whose objectives were to increase the number of AI users and improve livestock nutrition and feeding. However, inadequate funds allocated to this policy and high taxation level for vet pharmacies and feed industry limited its effectiveness.
- Subsidies on agricultural inputs were included in the "National agriculture policy" to enhance farm inputs' use and its efficiency. The limiting factors were inadequate financial resources allocated to the policy.
- Import and distribution of dairy cattle (with higher genetic potential) was included in the "Strategic and investment plan to strengthen the animal genetic improvement in Rwanda". The program aimed to increase the productivity of animal resources in a sustainable way and ensure that agriculture and livestock contribute to enhanced dietary diversity and quality at national and household levels. The following factors reduced policy effectiveness: taxes and licenses limiting semen imports for AI, low capacity of smallholders, lack of coordination, low involvement of the private sector.
- Tax exemption on importation of agricultural equipment and machinery in the "Agricultural Mechanization Strategy for Rwanda" policy, whose objectives were to increase the use of modern agricultural technologies and facilitate farmers in shifting from subsistence to commercial agriculture. However, available funds constrained effective policy implementation.

Extension services: They serve as focal points for facilitation and information related to the market, inputs, credits, and producer coordination. Options for extension delivery methods are becoming more pluralistic with the widespread use of mobile phones and information and communication technology (ICT). An extension communication system was built to allow direct feedback from extension workers to farmers for questions and queries. In addition, farmers can obtain information from different government institutions, also at decentralized level. This enabled farmers to access information on inputs markets available in the area. At the sector level, the government organizes savings and credit cooperatives (SACCO) which assisted farmers in obtaining loans for their business though micro-finance options. However, smallholders' access to extension advisory services was constrained by the exclusive availability of public extension agents and resources which were limited in size and scope.

Markets and other Institutions: Cooperatives, traders, and individuals were the most important marketing channels. However, half of sampled households had inadequate market access and were constrained by low and unstable farm-gate prices. The National Agricultural Export Development Board supports stakeholders' activities to process and export agricultural and livestock products. An exemption from taxation for selected agricultural inputs and equipment is an instrument established to enable and encourage the private sector to invest in agriculture.

Value chain development strategies

MAP members identified the following barriers to the adoption of *Brachiaria* forage in Rwanda despite its positive traits and benefits in terms of enhanced forage supply: lack of information on forage grasses, shortage of land for forage production, and lack of available seed material. The strategies suggested to overcome challenges included practical trainings, and on-farm demonstrations and trials that could mitigate the lack of technical and technological know-how, establishing a hub model for selling forage and exploiting cropping niches (e.g., under banana plantation) to overcome the issue of land shortage. Policies are required to support productivity enhancement through the increase in the available improved dairy cattle breeds. This can be reached by expanding the number of importers and streamlining the procedures for obtaining import licenses. Cooperatives might effectively provide both upstream and downstream services, facilitating access to input markets (fertilizers, credit) and training and serving as aggregators and quality promoters. The results from the Theory of Change exercise applied to the case study area are reported in Figure 9.3.

Discussion

The introduction of *Brachiaria* forage into the current farming systems of Kenya and Rwanda may generate ecological, nutritional, and socio-economic benefits

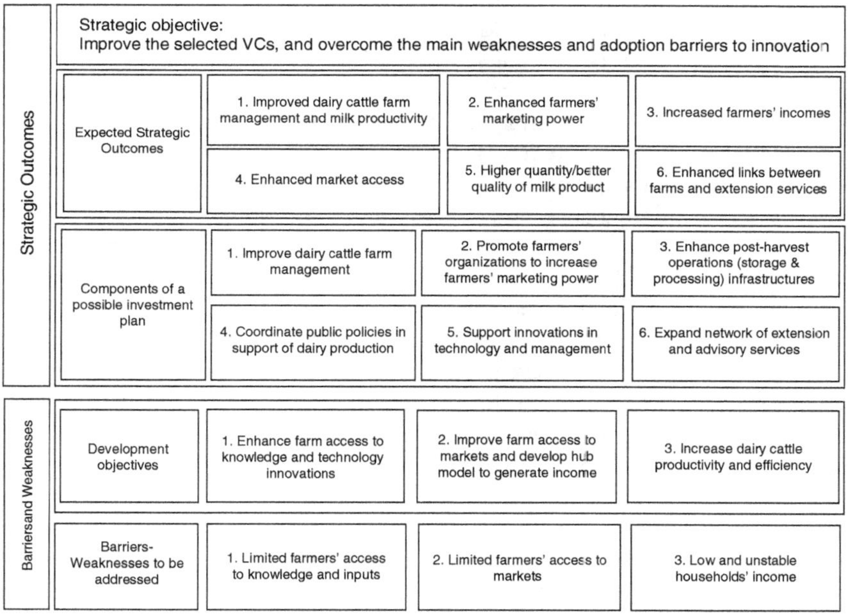

Strategic objective: Improve the selected VCs, and overcome the main weaknesses and adoption barriers to innovation			
Strategic Outcomes — Expected Strategic Outcomes	1. Improved dairy cattle farm management and milk productivity	2. Enhanced farmers' marketing power	3. Increased farmers' incomes
	4. Enhanced market access	5. Higher quantity/better quality of milk product	6. Enhanced links between farms and extension services
Components of a possible investment plan	1. Improve dairy cattle farm management	2. Promote farmers' organizations to increase farmers' marketing power	3. Enhance post-harvest operations (storage & processing) infrastructures
	4. Coordinate public policies in support of dairy production	5. Support innovations in technology and management	6. Expand network of extension and advisory services
Barriers and Weaknesses — Development objectives	1. Enhance farm access to knowledge and technology innovations	2. Improve farm access to markets and develop hub model to generate income	3. Increase dairy cattle productivity and efficiency
Barriers-Weaknesses to be addressed	1. Limited farmers' access to knowledge and inputs	2. Limited farmers' access to markets	3. Low and unstable households' income

Figure 9.3 Theory of Change: introducing *Brachiaria* into the dairy VC in Nyamagabe (Rwanda) (authors' own elaboration).

along the VC. For example, *Brachiaria* cultivation fostered higher milk yields, expanded product flow along the chain, and improved dairy farmers' incomes. Thanks to the promising market opportunities and consumers' demand for a wide range of dairy products. Also, *Brachiaria* grass can be introduced as an intercrop or border crop, and on marginal lands, being able to survive in poor soils with low nitrogen and phosphorus contents with evident positive externalities in the form of enhanced soil fertility and climate resilience.

However, smallholders' adoption of *Brachiaria* was constrained by limited access to VCs' opportunities, seed material, including value addition and transformation. Small farmers often operated with limited knowledge and capacity, and in a context of poor infrastructures and weak access to technology and knowledge services. Also, they cultivated small land parcels and could not introduce forage production due to the need to prioritize land use for crop production and food security purposes. Milk was undersupplied and economies of scale could be introduced along the chain, with efficiency losses for all operators.

In both case studies, strategies to overcome adoption barriers included information dissemination, demonstration, and on-farm trials to motivate new farmers to uptake forage cultivation, coupled with investments to enhance availability of forage seeds as well as suitable land areas for forage production (e.g., through irrigation). Actions to improve coordination along the chain may lead to more efficient dairy VCs.

In this context, multi-actor initiatives as MAPS have the potential to be a forum to enhance the diffusion of information and knowledge as well as coordination along VC actors, with benefits for all the participants. Such platforms face the problem from a multi-stakeholder point of view, and can identify suitable development strategies including options to harmonize institutions and agricultural policies to facilitate diffusion of agricultural innovation.

Our findings confirmed results from other studies available in the literature according to which the low innovations adoption by smallholders was influenced by farm size, farmer's education status, institutional assets, marketing possibilities, and profits (e.g., Kangogi et al., 2021). The effect of households' physical assets on technology innovation adoption was positive, due to households' improved management capacity (Mwangi and Kariuki, 2015). Access to knowledge is also a critical factor for adoption. For example, Obi and Maya (2021) showed that awareness creation targeting remote rural areas as well as institutions to ease farmers' access to information can contribute to higher adoption rates. Information access and association membership positively influenced technology adoption and innovation (Chowdhury et al., 2014).

Some limitations to our findings exist. Farmer entrepreneurship plays an important role in influencing adoption decisions of smallholders. Mizik (2021) showed that small-scale farmers consider the length of the payback period when they decide on any adoption of climate-smart agricultural practices. One way is to compensate them for providing environmental benefits, which is still not an option in the case study areas. Also, aspects related to drivers of coordination, cooperation, and institutional transformation processes, as well as to economic incentives to attract spontaneous participation of VC stakeholders deserve further investigation.

Conclusions

The case studies of Kenya and Rwanda presented in this chapter demonstrated that the diffusion of climate-friendly and resilient forage grasses such as *Brachiaria* offered promising results and demonstrated how technology innovations can transform current systems into CNRFS. However, investments are to be made to improve availability and on-farm access to forage grass seed material and improve fodder and dairy-cattle management. At the same time, technology and institutional interventions in off-farm VC segments (marketing, processing, storage, standards regulation) are required to capitalize the expected benefits deriving from on-farm innovations. Despite challenges, the sub-Saharan Africa is slowly becoming a competitive marketplace for agri-products.

Smallholder production systems must enhance their productivity in a more resilient way to respond to the increasing food demand in the context of climate change. As in other sustainability transitions, innovations in the technology as well as in the institutional settings play a critical role. Indeed, adoption of innovative technology to increase production efficiency and transform farming systems toward CNRFS will not be possible without farmers' access to properly functioning institutions, including effective information and knowledge systems, timely delivery of modern input technologies, and market access.

In this context, existing policies and institutions operating in the African agri-food system should be harmonized, along with an effective governance for multi-stakeholder VCs. The development of stakeholders' platforms – such as MAPs – represents an institutional innovation which could respond to such demand. Other studies have also shown that MAPs play an increasing role in scaling up innovations in agricultural systems (Barzola et al., 2020). In the two case studies, MAPs identified specific strategies to develop the VC in a coordinated manner. This included structuring the public-private EASs in support of the development of professional capacities and skills of extension workers; supporting cooperatives to enhance smallholders' participation in the VC, including their access to knowledge and inputs; improving regulations for license import of technical inputs for animal production; setting adequate hygienic standards related to milk commercialization; promoting public-private-producer partnerships on information and knowledge management; introducing labor market policies to lift the labor scarcity constraint and ease the adoption of labor-consuming innovation technologies. MAPs can provide a conducive entry point for smallholders' linkage with markets, especially those requiring assurances that adequate volumes of commodities can be traded. They will also play a key role in improving smallholder farmers' innovation skills and designing entrepreneurial agribusiness models, which could be replicated to different VCs and upscaled to national and regional markets.

Notes

1 The case studies refer to the activities conducted within the H2020 InnovAfrica project (www.innovafrica.eu) funded by the EU (Grant agreement no 727201 and call SFS-42-2016).
2 US$1 is equal to 1,183 RWF.

References

Agriculture for Impact (AFI) (2017) 'Agriculture for Impact: Agriculture Value chains'. Available at: http://ag4impact.org/sid/socio-economic-intensification/building-social-capital/agricultural-value-chains/

Barrett, C.B. (2008) 'Smallholder Market Participation: Concepts and Evidence from Eastern and Southern Africa', *Food Policy*, vol. 33, no. 4, pp. 299–317.https://doi.org/10.1016/j.foodpol.2007.10.005

Barzola Iza, C.L., Dentoni, D. and Omta, O.S.W.F. (2020) 'The Influence of Multi-stakeholder Platforms on Farmers' Innovation and Rural Development in Emerging Economies: A Systematic Literature Review', *Journal of Agribusiness in Developing and Emerging Economies*, vol. 10, no. 1, pp. 13–39. https://doi.org/10.1108/JADEE-12-2018-0182

Birner, R., Davis, K., Pender, J., Nkonya, E., Anandajayasekeram, P., Ekboir, J. and Cohen, M. (2009) 'From Best Practice to Best Fit: A Framework for Designing and Analysing Pluralistic Agricultural Advisory Services Worldwide'. *Journal of Agricultural Education and Extension*, vol. 15, no. 4, pp. 341–355. https://doi.org/10.1080/13892240903309595

Branca, G., Cacchiarelli, L., Haug, R. and Sorrentino, A. (2022) 'Promoting Sustainable Change of Smallholders' Agriculture in Africa: Policy and Institutional Implications from a Socio-economic Cross-country Comparative Analysis'. Available at: https://www.frontiersin.org/articles/10.3389/fsufs.2022.738267/full

Branca, G. and Perelli, C. (2020) "Clearing the Air': Common Drivers of Climate-smart Smallholder Food Production in Eastern and Southern Africa'. Available at: https://www.sciencedirect.com/science/article/pii/S0959652620319478?via%3Dihub

Branca, G., Cacchiarelli, L., Maltsoglou, I., Rincon, L., Sorrentino, A. and Valle, S., (2016) 'Profits Versus Jobs: Evaluating Alternative Biofuel Value-chains in Tanzania', *Land Use Policy*, vol. 57, pp. 229–240. https://doi.org/10.1016/j.landusepol.2016.05.014.

Branca, G., McCarthy, N., Lipper, L. and Jolejole, M. C. (2013) 'Food Security, Climate Change and Sustainable Land Management. A Review', *Agronomy for sustainable development*, vol. 33, no. 4, pp. 635–650.

Cafer, A. M. and Rikoon, J. S. (2018) 'Adoption of New Technologies by Smallholder Farmers: the Contributions of Extension, Research Institutes, Cooperatives, and Access to Cash for Improving Tef Production in Ethiopia'. *Agriculture and Human Values*, vol. 35, no. 3, pp. 685–699. https://doi.org/10.1007/s10460-018-9865-5

Chowdhury, A. H., Hambly Odame, H. and Leeuwis, C. (2014) 'Transforming the Roles of a Public Extension Agency to Strengthen Innovation: Lessons From the National Agricultural Extension Project in Bangladesh', *The Journal of Agricultural Education and Extension*, vol. 20, no. 1, pp. 7–25. https://doi.org/10.1080/1389224X.2013.803990

Clinton Global Initiative (CGI) (2016) 'Engaging Smallholder Farmers in Value Chains: Emerging Lessons'. Available at: https://www.clintonfoundation.org/sites/default/files/cgi_smallholder_report_final.pdf

Cronin, E., Fosselle, S., Rogge, E. and Home, R. (2021) 'An Analytical Framework to Study Multi-Actor Partnerships Engaged in Interactive Innovation Processes in the Agriculture, Forestry, and Rural Development Sector', *Sustainability*, vol. 13(11), pp. 6428.

de Oca Munguia, O.M. and Llewellyn, R. (2020) 'The Adopters Versus the Technology: Which Matters More When Predicting or Explaining Adoption?' *Applied Economic Perspectives and Policy*, vol. 42, pp. 80–91. doi:10.1002/aepp.13007.

Djikeng A., Rao, I.M., Njarui D., Mutimura, M., Caradus, J., Ghimire, S.R., Johnson, L., Cardoso, J.A., Ahonsi, M. and Kelemu, S. (2014) 'Climate-smart Brachiaria Grasses for Improving Livestock Production in East Africa'. *Trop Grassl-Forrajes Trop*, vol. 2, no. 1, pp. 38–39.

Ghimire, S., Njarui, D., Mutimura, M., Cardoso, J., Jonhson, L., Gichangi, E., Teasdale, S., Odokanyero, K., Caradus, J., Rao, I. and Djikeng, A, (2015) 'Climate Smart Brachiaria for Improving Livestock Production in East Africa: Emerging Opportunities', In: Vijay D. Srivastava, M.K., Gupta, C.K., Malaviya, D.R., Roy, M.M., Mahanta, S.K., Singh, J.B., Maity, A. and Ghosh, P.K. (eds.) *Sustainable Use of Grasslands Resources for Forage Production, Biodiversity and Environmental Protection.* Proceedings of 23rd International Grassland Congress, Range Management Society of India, pp. 361–370.

Hall, A.J. (2012) 'Partnerships in Agricultural Innovation: Who Puts Them Together and Are They Enough?' Available at: https://www.researchgate.net/publication/316736798_Partnerships_in_agricultural_innovation_Who_puts_them_together_and_are_they_enough

Haug, R., Nchimbi-Msolla, S., Murage, A., Moeletsi, M., Magalasi, M., Mutimura, M. and Westengen, O. T. (2021) 'From Policy Promises to Result through Innovation in African Agriculture?', *World*, vol. 2, no. 2, pp. 253–266.

Hermans, L. M., Haasnoot, M., ter Maat, J. and Kwakkel, J. H. (2017) 'Designing Monitoring Arrangements for Collaborative Learning about Adaptation Pathways,' *Environmental Science & Policy*, vol. 69, pp. 29–38. https://doi.org/10.1016/j.envsci.2016.12.005

Huffman, W.E., (2020) 'Human Capital and Adoption of Innovations: Policy Implications', *Applied Economic Perspectives and Policy*. vol. 42, no. 1, pp. 92–99. https://doi.org/10.1002/aepp.13010.

Husen, N. A., Loos, T. K. and Siddig, K. H. (2017) 'Social Capital and Agricultural Technology Adoption Among Ethiopian Farmers', *American Journal of Rural Development*, vol. 5, no. 3, pp. 65–72. https://doi.org/10.12691/ajrd-5-3-2

IPCC, 2014. AR5 Climate Change 2014: Impacts, Adaptation, and Vulnerability-Working Group II Contribution to the Fifth Assessment Report of the Intergovernmental Panel on Climate Change. https://www.ipcc.ch/report/ar5/wg2/

Kangogi, D., Dentoni, D. and Bijman, J. (2021) 'Adoption of Climate-smart Agriculture among Smallholder Farmers: Does Farmer Entrepreneurship Matter?' Available at: https://www.sciencedirect.com/science/article/pii/S0264837721003896

Kassie, M., Teklewold, H., Jaleta, M., Marenya, P. and Erenstein, O. (2015) 'Understanding the Adoption of a Portfolio of Sustainable Intensification Practices in Eastern and Southern Africa', *Land Use Policy*, vol. 42, pp. 400–411. https://doi.org/10.1016/j.landusepol.2014.08.016

Kolk, A., van Tulder, R. and Kostwinder, E. (2008) 'Business and Partnerships for Development', *European Management Journal*, vol. 26, no. 4, pp. 262–273.

Kosec, K. and Resnick, D. (2019) *Governance: Making Institutions work for Rural Revitalization.* In Global Food Policy Report, pp. 68–77. Washington, DC: International Food Policy Research Institute.

Kürschner, E., Baumert, D., Plastrotmann, C., Poppe A.K., Riesinger, K. and Ziesemer S. (2016) 'Market Access for Smallholder Rice Producers in the Philippines'. Available at: http://edoc.hu-berlin.de/series/sle/264/PDF/264.pdf

Li, L., Cao, R., Wei, K., Wang, W. and Chen, L. (2019) 'Adapting Climate Change Challenge: A New Vulnerability Assessment Framework from The Global Perspective', *Journal of Cleaning Product*, vol. 217, p. 216e224. https://doi.org/10.1016/j.jclepro.2019.01.162

Lybbert, T.J. and Sumner, D.A. (2012) 'Agricultural Technologies for Climate Change in Developing Countries: Policy Options for Innovation and Technology Diffusion', *Food Policy*, vol. 37, no. 1, pp. 114–123. https://doi.org/10.1016/j.foodpol.2011.11.001

Markelova, H., Meinzen-Dick, R.S., Hellin, J. and Dohrn, S. (2009) 'Collective Action for Smallholder Market Access', *Food Policy*, vol. 34, no. 1, pp. 1–7.

Mizik, T. (2021) 'Climate-Smart Agriculture on Small-Scale Farms: A Systematic Literature Review', *Agronomy*, vol. 11, pp. 1096. https://doi.org/10.3390/agronomy 1106109

Montefrio, M. J. F. and Dressler, W. H. (2019) 'Declining Food Security in a Philippine Oil Palm Frontier: The Changing Role of Cooperatives', *Development and Change*, vol. 50, no. 5, pp. 1342–1372. https://doi.org/10.1111/dech.12443

Mutimura, M. and Ghimire, S. (2020) 'Brachiaria Grass for Sustainable Livestock Production in Rwanda Under Climate Change', *Handbook of Climate Change Management: Research, Leadership, Transformation*, pp. 1–17.

Mutenje, M., Kankwamba, H., Mangisonib, J. and Kassie, M. (2016) 'Agricultural Innovations and Food Security in Malawi: Gender Dynamics, Institutions and Market Implications', *Technological Forecasting and Social Change*, vol. 103, pp. 240–248. https://doi.org/10.1016/j.techfore.2015.10.004

Mwangi, M. and Kariuki, S. (2015) 'Factors Determining Adoption of New Agricultural Technology by Smallholder Farmers in Developing Countries', *Journal of Economics and Sustainable Development*, vol. 6, no. 5, pp. 208–216. ISSN 2222-2855

Nagothu, U.S. (2015) 'The Future of Food Security: Summary and Recommendations', In: Nagothu, U.S. (ed.) *Food Security and Development: Country Case Studies*. London: Routledge, p. 274.

Nagothu, U.S. (2018) *Agricultural Development and Sustainable intensification: Technology and Policy Challenges in the Face of Climate Change*. London: Routledge. ISBN 9781138300590.

Njarui, D.M.G., Gatheru, M. and Ghimire, S. R. (2020) 'Brachiaria Grass for Climate Resilient and Sustainable Livestock Production in Kenya', In: Leal Filho, W. et al. (eds.) *African Handbook of Climate Change Adaptation*. Cham: Springer Nature, pp. 1–22.

Njarui, D.M.G., Gichangi, E.M., Ghimire, S.R. and Muinga, R.W. (2016) *Climate Smart Brachiaria Grass for Improving Livestock Production in East Africa – Kenya Experiences*. Nairobi: Kenya Agricultural and Livestock Research Organization, p. 271. https://cgspace.cgiar.org/handle/10568/79797

Norton, G.W. and Alwang, J. (2020) 'Changes in Agricultural Extension and Implications for Farmer Adoption of New Practices', *Applied Economic Perspectives and Policy*. vol. 42, no. 1, pp. 8–20. https://doi.org/10.1002/aepp.13008

Obi, A. and Maya, O. (2021) 'Innovative Climate-Smart Agriculture (CSA) Practices in the Smallholder Farming System of South Africa'. Available at: https://www.mdpi.com/2071-1050/13/12/6848

Poulton, C., Kydd, J. and Dorward, A. (2006). Overcoming Market Constraints on Pro-Poor Agricultural Growth in Sub-Saharan Africa. *Development Policy Review*, vol. 24, no. 3, pp. 243–277.

Raj R. and Nagothu, U.S. (2016) 'Gendered Adaptation to Climate Change in Canal-irrigated Agroecosystems'. In: Nagothu, U.S. (ed.) *Climate Change and Agricultural Development Improving Resilience through Climate Smart Agriculture, Agroecology and Conservation*. New York: Routledge, pp. 259–278, ISBN: 978-1-138-92227-3.

Rockstrom, J. and Barron J. (2007) 'Water Productivity in Rainfed Systems: Overview of Challenges and Analysis of Opportunities in Water Scarcity Prone Savannahs', *Irrigation Science*, vol. 25, no. 3, pp. 299–311.

Scialabba, N. E. H. and Müller-Lindenlauf, M. (2010) 'Organic Agriculture and Climate Change', *Renewable Agriculture and Food Systems*, vol. 25, no. 2, pp. 158–169.

Scopel, E., Da Silva, F. A., Corbeels, M., Affholder, F. and Maraux, F. (2004) 'Modelling Crop Residue Mulching Effects on Water Use and Production of Maize Under Semi-arid and Humid Tropical Conditions', *Agronomie*, vol. 24, no. 6–7, pp. 383–395.

Verderosa, D. (2021) 'New Hub Promotes Farmer Producer Organizations in India'. Available at: https://news.cornell.edu/stories/2021/09/new-hub-promotes-farmer-producer-organizations-india

Vohland, K. and Barry, B. (2009) 'A Review of In Situ Rainwater Harvesting (RWH) Practices Modifying Landscape Functions in African Drylands', *Agriculture, Ecosystems & Environment*, vol. 131, no. 3–4, pp. 119–127.

Wiggins, S. and Keats, S. (2013) 'Leaping and Learning: Linking Smallholders to Markets'. Available at: https://workspace.imperial.ac.uk/africanagriculturaldevelopment/Public/LeapingandLearning_FINAL.pdf

Wijk, J., Vellema S. and van Wijk, J. (2010) *Institutions, Partnerships and Institutional Change: Towards a Theoretical Framework*. No. 009. Partnerships Resource Centre.

Wiyo, K. A., Kasomekera, Z. M. and Feyen, J. (2000) 'Effect of Tied Ridging on Soil Water Status of a Maize Crop Under Malawi Conditions', *Agricultural Water Management*, vol. 45, no. 2, pp. 101–125. https://doi.org/10.1016/S0378-3774(99)00103-1

World Bank (2007) *World Development Report 2008: Agriculture for Development*. Washington, DC: World Bank. https://openknowledge.worldbank.org/handle/10986/5990

Woodhill, A.J. and van Vugt, S.M. (2011) 'Facilitating MSPs: a sustainable way of changing power relations?' In: *Guidebook for MSP Facilitation* (pp. 36–56). Deutsche Gesellschaft für Internationale Zusammenarbeit (GIZ) GmbH.

10 Enhancing and scaling climate-neutral and resilient farming systems

A summary and recommendations

Udaya Sekhar Nagothu

Introduction

Since the early 2000s, the debate about the carbon neutrality concept and its role in managing the risks and reducing the vulnerabilities in the land and the food system has drawn the attention of scientific community and policy makers (Becker et al., 2020). At the same time, the awareness and demand for sustainable and green products is gradually increasing among urban consumers and society at large. Climate-neutral and resilient farming systems (CNRFS) must gradually replace the current systems that are intensive, polluting and unsustainable (IPCC, 2022). In fact, CNRFS are not only beneficial to the climate but could also be motivating to the farmers if incentivized for the carbon credits they generate in the process of adopting them. Studies show that farms can offer a huge potential for reducing methane emissions and increasing carbon sequestration, and thereby support global commitment to reduce greenhouse gases (GHGs) (OECD, 2019; Lehner and Rosenborg, 2021).

However, science and technology alone cannot stop global warming. It must be supported by appropriate policies and regulatory systems, political will, collective stakeholder responsibility, wider support from the public at large, adequate investments to promote systematic implementation and scaling of climate-neutral (or carbon-neutral) and sustainable production systems (UNFCCC, 2019a). There is no 'one-size-fits-all' solution to address climate crisis. We need a mosaic of climate adaptation and mitigation options that will suit different situations, where the environmental, social and economic contexts and vulnerabilities are duly considered (OECD, 2021a). The measures, preferably, should include the old, current and new ones, the ecosystem-based practices that farmers practiced, rethinking the use of agrochemicals and replacing these with bio-based solutions, including precision-based applications of inputs (IPCC, 2022). In a recent 'Farming for the Future' report, the authors emphasized that achieving climate neutrality in agriculture is not only about climate but aiming toward a sustainable food production regime based on principles of resource efficiency, productivity and farm profitability (SIANI, 2020a). The sustainable food systems approach aims at strengthening the entire value chain, changing not only the way we produce but also how we consume and avoid food wastage.

DOI: 10.4324/9781003273172-10

The Glasgow COP26 summit agreement to reduce GHG emissions is encouraging (UNEP, 2021). However, the deal in its current form may not be enough to limit global warming to 1.5°C over pre-industrial levels. At the same time, the roadmap for the Nationally Determined Contributions (NDCs) of several countries falls short of the expectations, according to the analysis of the United Nations Environment Programme (UNEP, 2021). Though climate-related agreements were made in the past, countries have not managed to put them into action due to various constraints. Only if countries or regions can see climate crises as an opportunity to innovate, create jobs, and cooperate with each other to reduce climate instability and human insecurity, can we be hopeful about the future climate action (OECD, 2009)? Initiatives like the Horizon Europe program and the EU Farm to Fork Strategy are good examples of how innovation and cooperation can help to drive climate action forward (EU, 2021). Using market and economic instruments and putting a price on the carbon and emission reductions in the agriculture sector would encourage farmers to adopt CNRFS.

The impact of COVID-19 on economy and food security and the prevailing geopolitical conflicts in the Asia-Pacific region, Europe and parts of Africa will make it even more challenging for governments to prioritize, cooperate and support the global climate agenda (OECD, 2021b). The concern is not only about developing nations but also about developed countries, which are the major sources of GHG emissions, not fulfilling their commitments and making adequate investments. It is crucial that the major GHG emitters, in terms of the total emissions as well as the per capita emissions, must come on board and cooperate for the global climate good (UNCTAD, 2021). Advocacy and diplomacy using the existing international bodies and forums such as the UNEP, the EU, the IPCC, the COP summits and other platforms must be continuously pursued to engage countries that are not willing to take climate action seriously (IPCC, 2021).

The IPCC report provided a more detailed regional assessment of impact of climate change, concluding that it is already affecting every region on the Earth (IPCC, 2021). The growing scientific evidence about climate change must be taken seriously by politicians and bureaucrats. Their support now to promote CNRFS to improve adaptation and reduce GHG emissions will define the future of our climate action. Though the limited funding opportunities in developing countries will force governments to follow the economic agenda rather than invest in climate action, there is still hope. One way to address this challenge is by ensuring that development work is 'climate proofed' and that climate action is to be development oriented (SIPRI, 2019). In this way, the environmental and climate priorities no longer need to be put secondary on the development agenda of the countries.

Optimism about climate change mitigation

Keywords for driving climate action are appropriate technology innovations and sharing the knowledge, multi-stakeholder cooperation, training and capacity building, financing and transparency in reporting and supportive policies

(UNFCCC, 2019b). There are positive initiatives already in place, from international to national actions being undertaken, accompanied by increased commitment to investments. Though it is still not adequate, we need more rapid transformation. The EU has been in the forefront when it comes to climate action supportive policy and implementation (Delbeke and Vis, 2016). For example, the EU Green Deal and the Common Agricultural Policy (CAP) strategic plans have set time-bound targets to achieve climate neutrality in agricultural production by 2050 (European Commission, 2021; EEA, 2021). Such political and economic commitments to innovative action, research and development is the key to drive the climate action forward. The initiatives also put emphasis on sharing knowledge and networking among stakeholders for promoting upscaling of the innovations. Similar initiatives in the energy sector in India, China and South Africa are driving action toward renewable energy transformation by large-scale investments in solar and wind power (UNFCCC, 2017).

Despite the disagreements, some of the commitments made at the 2021 Glasgow COP26 summit are promising (United Nations, 2021), including (i) the US- and EU-led commitment by over 100 countries to slash 30% methane emissions by 2030; (ii) promoting nature-based solutions to reduce GHG emissions and reduce global warming; and (iii) ending deforestation by 2030 as pledged by 120 countries. To achieve the set targets for the three agreements, it is vitally important to ensure technology sharing, ensure genuine cooperation among the countries and provide the climate finance pledged by developed countries to support carbon economy for the benefit of millions of smallholders involved in agriculture and forestry sectors in developing countries. Integrating gender and youth in the process, especially, their role in the intended NDCs, cannot be ignored in the process (The Commonwealth, 2021). Any gender mainstreaming attempt must be accompanied by planning and instigating gender response budgeting and follow-up. There is growing recognition of the role of gender in climate action (IISD, 2019a). The year 2030 is also the year when the Sustainable Development Goals (SDGs) will have to be achieved, especially, to curb poverty, achieve Zero Hunger and secure the planet's future (UN, 2020). In this regard, climate action investments could be seen as an opportunity to address the entire SDG agenda.

Transition to carbon neutrality in agriculture

Transition to carbon neutrality in small-scale agriculture is not an easy goal with governments facing various challenges already due to economic and climate crisis as discussed in Chapter 1. We need to make a start somewhere, by supporting changes in farming practices at different levels. Practices that can effectively improve efficiency of soil, water, pest and overall crop management, and at the same time improve adaptation and mitigation to climate change by reducing GHG emissions and increasing carbon in the soils (IGI, 2021). Simple and low-cost carbon-neutral farming solutions can make a large impact in the agriculture sector. Disparate chapters in this book demonstrated the climate adaptation and mitigation benefits of precision soil, nutrient and water resources management, integrated pest management, climate-resilient systems of sustainable rice intensification, direct sowing of rice and agroecology-based practices that provide multiple benefits that are suitable for implementation on small farms with low investments. The focus of the book has been mostly on rice, drawing evidence from authors' own experiences, multi-country and -stakeholder action research and development initiatives, observed field data and results from other relevant studies.

As majority of the farms globally are small scale in nature, smallholders practicing CNRFS must be incentivized properly for their efforts to reduce emissions from the agriculture sector. My own work and experiences during the last 25 years in Asia and Africa and other similar studies have shown that investments and support for climate-neutral agricultural production for smallholders are limited (Nagothu et al., 2018; IFAD, 2021). For that matter, support for smallholders to reduce emissions is still absent in many countries and promoting the adoption of CNRFS on the ground and at scale is far from ideal in the current reality. For example, in African countries such as Malawi and Zambia, incentives in the agriculture sector are mostly restricted to fertilizer subsidies for maize production alone as discussed in Chapter 8. In these countries, the government priority is to ensure food security primarily by increasing maize production. This is similar in many other countries in Africa, where governments are constrained with resources, and agriculture is based on subsistence production to meet farm households' own consumption. In South and Southeast Asia, the situation is different, with governments subsidizing the farmers for the use of seeds, water, fertilizers and electricity (ADB, 2021). Gradually, we are seeing regulations and incentives being introduced in countries such as India, Vietnam, Thailand and Japan for efficient use of water and reduced use of chemicals in agriculture, to improve overall efficiency and productivity, in addition to the promotion of more environment friendly production systems, including organic farming.

For successful climate mitigation, smallholders need timely access to technology, capacity building, affordable and quality seeds, market access, fair prices for their products and economic incentives for implementing the CNRFS. Interventions in agriculture to increase climate adaptation and mitigation do not need to be linked to high technology in all cases as demonstrated in the book.

The final chapter begins with an introduction and briefing on the current opportunities and constraints for climate action, followed by a section with main messages from each of the book chapters. The chapter summarizes factors that are necessary for promoting CNRFS based on results and discussion from the book chapters. The focus is on technology, investments, stakeholder cooperation and policy options necessary for enhancing and scaling the adoption of CNRFS.

Brief summaries of chapters

Chapter 1 presented an overview of the current climate crisis, major sources of GHG emissions, and impacts from the agriculture sector contributing to global warming. Further, the chapter discussed the challenges in reducing GHG emissions from the agriculture sector and the major changes required in the agriculture sector if the impact due to climate change is to be limited to 1.5°C target. The solutions to overcome the challenges must be designed for different agroecological settings, and implemented at different scales, both for developed and developing countries, for large- and small-scale farms, and should be sustainable, environmentally, socially and economically. The chapter discussed the possible pathways for a transition toward sustainable CNRFS, and toward the end provided a brief outline of the book.

In Chapter 2, the authors focused on the importance of *precision-based soil and nutrient management* practices tested on rice farms in the eastern part of India and the potential for reducing GHG emissions and increasing soil organic carbon. This is highly relevant for countries such as India, Vietnam, Myanmar, Bangladesh and Thailand with large areas under rice production where the use of excess amounts of fertilizer and other chemicals, especially the nitrogen fertilizer, is a serious problem for the environment and health of people. The chapter has shown the importance and benefits of using tools ranging from the simple leaf color chart to innovative digital tools and their relevance to improve nutrient use efficiency. The chapter toward the end provided guidelines and policy recommendations for upscaling precision soil and nutrient management in rice systems and other related food crops.

The importance of water management and its role in mitigation of GHG emissions in the Asian region was the focus in Chapter 3. Efficient water management is required to suppress the emission of methane from paddy fields, which is the major source of methane in agriculture. In addition, water saving is possible through improved water management as water resources are becoming scarce due to climate change and extreme weather. According to the authors, alternate wetting and drying (AWD) is a promising water management method for reducing methane emissions from paddy rice and at the same time improving water use efficiency. This is possible with the alteration in redox state of soil through AWD. Although its methane-reducing and water-saving effects were demonstrated by previous studies, AWD is not widely practiced in rice growing regions in Asia due to associated constraints. To implement and realize the benefits of AWD at a systemic level, it is necessary to manage the irrigation system collectively by farmers (e.g., water users' group). AWD is not possible without cooperation between farmer organizations and government agencies, which is necessary for joint monitoring of water release and supply. The chapter discussed in detail the practical AWD implementation through 'block-wise distribution' system of water supply managed by water users' group themselves in the paddy district of the Red River Delta area in Vietnam. Specifically, the effects on methane emission and rice yield, as well as the limitations for AWD were analyzed. In addition, the institutional and technical measures supporting organizational and environmental friendly agricultural water management necessary for upscaling AWD were discussed.

Pests and diseases are a serious challenge to crop production, which will be impacted by climate change and temperature fluctuations that can result in new outbreaks. Currently, chemical control dominates the pest management programs worldwide in agriculture. Chapter 4 provided a comprehensive review of various *integrated pest management* (IPM) measures in rice in combination with nature-based or agroecosystem-based interventions, physical, cultural and biological practices that provide additional benefits toward conserving biodiversity, ecosystem services and climate change mitigation. The chapter illustrated the importance of climate-smart and sustainable IPM and the potential ecological, economic and social sustainability impacts it can generate. The chapter discussed the relevance of threshold-based and spatially targeted application of pesticides supported by digital tools and models. A key factor to bridge gaps between scientific knowledge and practical implementation of IPM measures is the continuous involvement, training and co-designing of solutions with farmer communities and other stakeholders, using approaches such as the FAO *farmer field schools.* The chapter presented some examples of IPM from an ongoing field project in India and the multiple benefits they can generate in rice.

In Chapter 5, the authors analyzed the role of agroecological based climate-smart rice farming systems with focus on *system of rice intensification (SRI)* in climate adaptation and mitigation. Modern agriculture contributes significantly to climate change through methane production, with paddy rice production being a major source. However, there is a potential to reduce the negative impact through SRI. Methane emission from flooded rice fields can be reduced up to 50–70% by applying certain practices such as reduced density in planting combined with AWD irrigation cycles. At the same time, the SRI intervention can increase yield by encouraging aerobic soil conditions, improving soil health and stimulating root systems. The latter creates opportunity for additional carbon sequestration. However, there are disagreements within the scientific community regarding the extent of benefits of SRI.

Rice fields host, both above and below ground, a rich diversity of species, and are a source of multiple ecosystem goods and services that are vital for producing healthy plants and enhancing yield. Smallholder rice farmers, responsible for the bulk of global rice production, can play a major role as custodians and managers of the largest human-made wetlands and the biodiversity conserved within these agroecosystems. But to realize these benefits, a major transition must take place in rice paddy cultivation – moving away from the current predominantly extractive and exploitative models and shifting toward regenerative methods of agriculture production. Chapter 5 provided a conceptual framework for optimizing the management of ecosystem goods and services reflected in the biodiversity contained in healthy rice paddy fields. It described existing successful examples of rural communities empowered to appreciate and responsibly manage rice production landscapes and their *in situ* biodiversity. The chapter outlined the enabling environment toward a more *sustainable* intensification of rice production at landscape, national and global levels. Such a transformation aligns with the actions called for in the UN Decade of Ecosystem Restoration (2021–2030) and will assist in achieving the UN Sustainable Development Goals by 2030.

In Asian countries, rice is commonly grown by transplanting seedlings in puddled soils which is labor-, water- and energy-intensive practice. Moreover, the puddling operation deteriorates the soil physical structure, which adversely affects the performance of succeeding crops. The increasing water demand and labor scarcity needs a major shift from conventional puddled transplanted rice to direct seeding of rice (DSR), especially in the irrigated ecosystems. Chapter 6 provided a summary of research findings from the case studies in India that showed significant benefits of another climate-smart rice system, namely the *direct seeded rice* (DSR), which gives positive outcomes compared to puddled transplanted rice in terms of (i) increased water productivity if managed properly; (ii) reduction in labor and production costs; and (iii) lower methane emissions. Nevertheless, there are challenges for adopting DSR, including poor weed control, availability of suitable climate-tolerant varieties for DSR, increased damage by soil pathogens, and nutrient disorders, especially nitrogen and micronutrients. Possible solutions to overcome these challenges that will make it easier for adoption by farmers were analyzed in Chapter 6. Observed field data and evidence from other studies under both dry and wet conditions from India and other previous studies were presented to support the solutions. The options for scaling up DSR supported by farmer trainings, accessibility to good quality seeds, availability and use of drum seeders and policy were discussed toward the end.

In Chapter 7, the authors discussed the impacts of current farming systems in Europe, which are in general intensive, highly specialized and mechanized, heavily dependent on external input and with no or very low crop rotation options. This type of farming system model has a negative environmental impact on water, soil and biodiversity as well as on GHG emissions. Fallow season and inappropriate crop residues management, with no rotation, low utilization of manure and organic fertilizers (due to absence of livestock farming in the area), and intensive soil laboring has led to high soil carbon losses. According to the authors, this could be reversed by adopting proper land management and shifting to organic production that can contribute to biodiversity and agroecosystem services improvement. The chapter shared experiences from rice cultivation in northern Italy, under improved soil and water management, and the mitigation potential through soil carbon storage enabling rice fields to act as carbon sink. The chapter presented main findings from their work on carbon-neutral farming solutions to reduce GHG emissions (including nitrous oxide emissions from fertilizers) and increase carbon sequestration in the soils. The chapter toward the end presented the EU agricultural policies and climate actions in support of the adoption of GHG mitigation solutions, which could be a good example for other countries.

Today's African farming systems are unsustainable, as they rely on high external inputs such as water, mineral fertilizers and pesticides to increase agricultural production. This has resulted in serious degradation of soils, water, biodiversity, ecosystem services, vulnerable to pests/diseases outbreaks and climate change. Chapter 8 provided a comprehensive review of literature pertaining to agroecological (AE) farming approaches/practices and knowledge driven from stakeholders' and scientific studies. The review identified the major drivers, barriers, gaps and opportunities of AE practices in the context of African farming systems. Experiences from Zambia with agroforestry practices in maize were shared in the chapter. Further, key ecological, social and economic indicators developed in the countries were also discussed. The chapter analyzed how the AE practices con-tribute to the reduction of GHG emissions and at the same time address the UN Sustainable Development Goals (SDGs), e.g., SDG 2 (*food and nutrition security*), SDG 12 (*sustainable food production and consumption*), SDG 13 (*climate action*) and SDG 15 (*life on land*).

Chapter 9 begins with a brief introduction followed by a conceptual framework showing the linkages and interactions between different institutional, market and policy factors affecting adoption of CNRFS in the agriculture sector. The chapter then discussed the barriers for adoption, which operate at various levels in the value chains (VCs). The role played by stakeholders (VC actors, farmers group, research, government agencies and donors) in the farmers' adoption and the dynamics and partnerships to be developed between different VC actors for upscaling CNRFS was analyzed. Experiences from case studies in Africa (Kenya and Rwanda) were shared demonstrating how strategies to overcome weaknesses and adoption barriers in the selected value chain together with the support of multi-actor partnerships. Toward the end, some concluding remarks and policy recommendations for upscaling CNRFS were provided.

The final chapter in the book summarized the key messages and recommen-dations from the disparate chapters. The chapter provided the challenges and at the same time an optimistic note about future climate actions citing some of the ongoing good initiatives. The possible technical, investment and policy options necessary for upscaling CNRFS were highlighted toward the end.

Good practices for climate-neutral and resilient agriculture production

To achieve climate-neutral and resilient agriculture production, the governments, together with relevant stakeholders, must facilitate and improve technical, organ-izational and investment solutions suitable for different agroecological settings (INRAE, 2021). A first step in the transition toward sustainable CNRFS will be to make a good baseline assessment of the current systems, including the strengths and weaknesses at field/landscape scale, interdependencies, drivers and barriers, and the opportunities as discussed in Chapter 1. A good baseline data and cor-responding indicators would be useful for measuring, monitoring and verification of impacts observed with the introduction of CNRFS. The baseline scenarios will be useful for the next stage of *co-designing* combinations of the most effective solutions, both old and new, that are of low cost and of low risk to smallholders, targeting soil, water and crop management that are sustainable and agroecosys-tem based (Dainese et al., 2019). In the process, it is important to have a close cooperation and active engagement with farmers and relevant value chain actors. Wherever possible business models for the specific agri-products must be devel-oped that will give smallholders better access to more rewarding markets.

Table 10.1 provides a list of potential options that are climate resilient and at the same time have the potential to reduce GHGs. It is the net GHG reduction and carbon sequestration that should be considered in the overall food system

Table 10.1 A list of farming systems to reduce GHGs, and improve resilience to climate change

Interventions	Practices leading to CNRFS	Chapter	Implications for climate action
Soil and nutrient management	Precision-based soil and nutrient management; agroecosystem-based measures; apps and digital tools supporting precision, farmer training, farmer incentives	2	Improving soil health, reducing chemical inputs, increasing soil organic carbon and reducing GHG emissions
Water management	Alternate wetting and drying in rice system of rice intensification Institutional support to AWD and SRI upscaling, participatory extension, farmer field schools	3, 5	Increasing water use efficiency and productivity, and reducing methane emissions
Integrated pest management	Habitat management, integrating agroecosystem/-nature-based measures; biological, cultural and physical control measures; training and capacity building of smallholders in IPM, including through farmer field schools	4	Reducing pest damage, reducing the use of chemicals, increasing yields, protecting biodiversity and ecosystem services
Cropping systems and climate-resilient varieties	Crop rotations, intercropping, agroforestry, system of rice intensification, direct seeding of rice; agro-forestry, climate-tolerant varieties (*drought and flood resistant, nitrogen fixing*)	5, 6, 7, 8	Increasing yields, reducing methane emissions, overall efficiency and adaptation
Institutional and policy support	Multi-actor engagement, value chains and VC actors' integration	9	Developing business models, strengthening value chains, improving access to markets and farmer income

Source: Authors' own compilation.

development. A particular farming practice, take for example AWD, may reduce methane but could lead to higher emissions of other GHGs, such and nitrous oxide, and other negative trade-offs. Hence, care must be taken to look at the overall environmental, economic and social impacts in totality of the new farming systems introduced. In theory, it is technically feasible for the agriculture sector to become close to carbon neutral relying on supply-side mitigation measures alone, although it depends on optimistic assumptions about the potential of soil carbon sequestration (OECD, 2019). Within the EU, a target to achieve climate neutrality in the agriculture sector has been set for the year 2050 (European Commission, 2021). How this can be translated into practice will depend on the changes at different scales and at individual farm level. The implications to set such objectives in developing countries would be different, as it would involve millions of small-scale farms that are scattered and difficult to reach and monitor. It will be a daunting task for the resource-constrained governments to bring a significant percentage of these farms to change their current farming practices and shift to CNRFS. Systematic and step-wise building of climate action initiatives starting with ecosystem or nature-based solutions will be possible with careful planning, investments and capacity building. When farmers see the economic incentives, upscaling becomes easier.

Integration and cooperation

Actions to address climate crisis, with specific efforts to reduce GHG emissions and global warming, need a *well-integrated multidisciplinary approach*. Though multidisciplinary research has been receiving recognition in climate action research in recent years, in practice, it is still a challenge to make scientists work across their disciplinary silos (Nagothu, 2015). International initiatives such as the EU Horizon Europe programs are putting emphasis on multidisciplinary or cross-disciplinary approaches as a necessity for acquiring funding, which makes it obligatory for scientists to cooperate and work beyond their disciplines (European Commission, 2021).

At the next level, integration is necessary between scientists and stakeholders, representing different sectors including relevant government agencies (agriculture, extension services, environmental agencies), farmer and business organizations. The science-policy linkage through *multi-actor engagement* is another important tool for addressing the climate crisis (Nagothu et al., 2018). It will help to overcome challenges in science-policy gaps that are mainly related to translation, communication, time lag, uncertainty and credibility (JPI Climate, 2021). Chapter 9 demonstrated the importance of multi-actor platforms that help in encouraging cooperation and thereby better implementation of CNRFS. One way is through integration of stakeholders into scientific research that will make the research outputs more relevant to address their needs. At the same time, scientists must develop solutions that are socially and politically acceptable and consider the trade-offs resulting from new technologies introduced to reduce GHGs and/or improve carbon sequestration (Nkiaka and Lovett, 2019). Thus, knowledge co-creation with active involvement of stakeholders can make science-policy dialogues more purposeful (IISD, 2019b). As discussed in Chapter 1, uncertainty in science increases skepticism among politicians that could lead to wrong and/or biased decision-making. Hence, scientists need to make efforts to provide the right scientific evidence at the right time with accuracy.

Adequate space or platforms for dialogue must be further strengthened and promoted where stakeholders can meet, exchange and discuss needs, challenges and options to address climate action. The science-policy interaction platforms will improve the chances to influence that scientific robustness is not compromised while formulating policies (Howarth and Painter, 2016). Such a dialogue and cooperation between scientists and policy makers becomes even more relevant, as countries are revising their new climate action plans to be included in their 'Nationally Determined Contributions' (NDCs). In the process, transparency and open dialogue among relevant actors will be in the interest of all parties (UNFCCC, 2019a,b). Cooperation and sharing knowledge among agencies, at national and global levels, must further improve for the benefit of the society at large (International Science Council, 2021; IPCC, 2022). This will be one of the key factors for promoting climate action globally and motivating the developing countries that do not have the access to mitigation technologies.

Technology options

There is a lot of focus on technological solutions to reduce GHGs and increase soil carbon sequestration. We also see a change even in developing countries where transformation of food systems is happening both on the production and consumption side (Reardon et al., 2019). Market and consumer preferences for organic and reduced carbon footprint are encouraging farmers to adopt innovative CNRFS.

Within agriculture, maintaining soil health is one of the most critical factors necessary for overall sustainable and carbon-neutral agricultural production. A first step toward improving soil health is to introduce regular and site-specific soil testing to assess the status and plan soil improvement measures accordingly. The 'Soil Health Card' scheme initiated in India is a good example of how to organize a nationwide soil testing campaign toward improving soil fertility (Government of India, 2015). In the process, such measures also aim at improving organic carbon content of soil and ecosystem services. As demonstrated in Chapter 2, there are measures such as *minimum or zero tillage, crop rotations with legumes, cover crops* and use of *organic mulching* that give multiple benefits. Zero tillage is now being promoted in India by CIP and has shown good results (CGIAR, 2020). Chapter 2 also demonstrated the impacts due to *precision application of nitrogen* fertilizer application that can significantly reduce the amount of chemical fertilizer used and lower GHG emissions. In this context, the use of sensors and digital tools is becoming popular for guiding farmers on proper timing and precision use and management of nitrogen application. A need-based application of fertilizers also reduces excess of fertilizers and input costs to farmers.

An international initiative '4 per 1000', launched at the 2015 Paris climate conference, showed that increasing soil carbon globally by a mere 0.4% annually could offset that year's new growth in carbon dioxide emissions from fossil fuels (Van der Pol, et al., 2021). Most carbon-farming techniques mirror age-old organic growing methods that have the capacity to contribute to a healthy soil system, for example, the use of cover crops and organic mulch (Barth, 2016). Cover crops are common in some countries, which help to capture carbon from the atmosphere and store it in soils, in addition to adding nutrients (WUR, 2019). These practices are termed as 'regenerative agriculture', which are low-cost solutions to increase soil organic carbon and are suitable for both small and large farms. Chapter 8 discusses the impacts of systems such as cereal-legume rotations and integrated cropping systems on GHG emissions and soil carbon sequestration. According to some studies, diverse cropping systems alone will be able to sequester soil organic carbon and increase biodiversity simultaneously (SIANI, 2020a).

Some of the common problems in agriculture including soil erosion, poor soil health, nutrient removal, and runoff can be controlled by practices that can also increase soil carbon simultaneously. Soil carbon should be seen as another farm output, and accordingly incentivizing farmers that help in increasing soil carbon will be necessary for reducing GHGs and improving soil health and productivity (van der Pol et al., 2021). Within the EU, financial support mechanisms already exist to support farmers to practice soil mitigation measures. Studies have shown that integrating improved farming practices including precision-based fertilizer application, supported by soil tests, reducing summer fallow frequencies and crop rotation of cereals with grain legumes, lowers carbon footprint in crops such as wheat at an average of $-256 \, kg \, CO_2 \, eq \, ha^{-1}$ per year (Gan et al., 2014). Therefore, it makes sense to include net-zero farming and carbon capture initiatives to promote climate mitigation into NDCs (IGI, 2021). Though in practice it may not be easy to reach net-zero farming levels, we should start somewhere.

One of the basic premises for an integrated approach is a good understanding of the agroecosystem, which will help in planning and developing most suitable measures to address the climate crisis. Be it integrated soil and nutrient management or pest management, the new measurers must be ecosystem based as well as climate smart. Developing climate-smart and sustainable IPM will be a necessary part of carbon-neutral agriculture as shown in Chapter 4. By reducing damage to crops from pests and diseases, farmers can sustain yields and thereby reduce pressure on more land for production. Chapter 4 also demonstrated the importance of IPM as the way forward to combat multiple and new invasive pest outbreaks in the future due to extreme weather events that are increasing. Some of the IPM measures are low cost but require additional efforts in terms of collective action where several agencies and farmers in the locality should cooperate for better results.

Within agriculture, improving water use efficiency and reducing methane emissions are the two main objectives that are possible through alternative climate-smart management practices such as the *Alternate Wetting and Drying* (AWD) irrigation which was the focus of Chapters 3. Improving the efficiency of water management in paddy rice will be one of the most important measures in agriculture to address, as paddy rice is a major contributor of methane. Any climate action would not be able to reach its targets without significant reduction of methane from rice and livestock sectors. The SRI and DSR systems (Chapters 5 and 6) have been investigated as potential climate mitigation measures in rice by several agencies in the past. But the results so far are mixed and still debated by scientists. These systems have their potential, benefits and challenges at the same time, depending on agroecosystem in which they are introduced, availability of suitable rice varieties, weed infestation problems, capacity of farmers, scalability and so on. Hence, more research and especially development initiatives need to be pursued to establish the credibility of these systems. The results presented in Chapters 3, 5 and 6 are thus another attempt in this direction, adding new knowledge and reasons to practice these systems in the major rice growing regions of Asia. The chapters demonstrated practical measures to overcome some of the barriers in upscaling.

New hybrids of food crops that can adapt to extreme weather, including floods and droughts, and simultaneously lower carbon footprint will be in demand in the future. In Chapter 6, the benefits of such varieties in rice are demonstrated for DSR. One of the constraints is to mass-produce the new seed varieties and make it accessible to smallholders in developing countries. Further, increasing overall farm productivity would imply a reduction in GHG emissions (Rural Hub, 2021). Also, DSR in dry fields, rather than transplanting as often done in flooded rice paddies, will help to significantly improve soil organic carbon if crop rotation is practiced with legumes. Thus, DSR could be another promising practice in rice that could play a significant role in climate mitigation as well as soil health promotion.

There is much emphasis these days on *nature-based solutions* to address the climate crisis, i.e., by taking advantage of nature to provide sustainable and cost-effective solutions to climate problems (SIANI 2020b). For example, *organic agriculture, agroforestry and other interventions* that can improve soil fertility, increase soil organic carbon and at the same reduce GHGs as discussed in Chapters 7 and 8. Agroforestry is common in many developing countries, practiced by smallholders, and a potential climate mitigation measure that can provide multiple advantages and additional income to farmers. Similarly, the importance of organic agriculture, and the consumer awareness for organic and locally grown products with reduced water and carbon footprint, is increasing gradually among segments of society, including the youngsters and urban based consumers.

Nationally Determined Contributions and climate action

An awareness about the multiple benefits of CNRFS discussed so far should be created among policy makers and planners preparing the NDCs (UNFCCC, 2019a,b). The achievements of NDCs will depend on how the countries plan, implement, follow up and communicate the progress with transparency. The NDC support program is helping developing countries to reduce GHGs (UNDP, 2021). It is not only investments *per se* that are necessary to drive the NDC goals, but also the prioritization of investments, technology sharing and open communication. It is crucial that some of the major emitters of GHGs follow their national commitments, including the USA, China, India, Brazil and others without which it will be difficult to persuade smaller nations. It is positive to note that more than 70% of the countries have submitted their NDCs by 2021, though most of them have not started to put them into action. More than 80% of the countries have agreed to use the international market mechanisms to achieve mitigation goals by trading carbon credits or offsets with other countries (WRI, 2021). The Article

6 of the Paris Climate Agreement establishes that Parties may elect to cooperate with other countries, in the process of implementing the NDCs. The Article thus legitimizes cooperation and opens opportunities for smaller nations that have funding constraints. Sectors such as agriculture and forestry in developing countries could benefit from this Article where carbon credits could be traded for changing agricultural and forestry practices not only to reduce emissions but also to conserve the tropical forests and biodiversity. Without adequate investments, it may be difficult to motivate farmers to adopt CNRFS.

Investment options

It is encouraging to know that most scenarios include the likelihood of agricultural emissions being reduced in the coming decades (Kingwell, 2021). Innovations in carbon-neutral agriculture and development that enable reduced agricultural emissions will be critical to lower the cost of achieving and sustaining carbon neutrality. The EU Horizon Europe program is a good example of how investments to support innovative research and development can integrate public and private sectors, farmer groups and government agencies to drive climate-neutral agenda forward (European Commission, 2021). Innovations that will be developed under this program will be made open for use by relevant stakeholders within and outside the EU. Similar initiatives outside the EU are necessary to engage with relevant actors, share knowledge and cooperate on climate action. International funding facilities such as the Climate Support Facility (CSF) seek to align the green economic recovery efforts with the national climate goals and climate-resilient strategies they adopt in different sectors (World Bank, 2021). Under the CSF, the new multi-donor trust fund, the NDC Support Facility (NDC-SF), was created specifically to facilitate the NDCs. Similarly, the NDC Invest and the Climate Investments Fund support climate action in the low- and middle-income Latin American and Caribbean countries (Fazekas et al., 2021).

The role of private sector, including philanthropic groups and individuals and corporate sector, will be important to secure adequate climate financing, as governments alone cannot provide the necessary resources for meeting the costs of mitigation in the agriculture sector. For example, the 'Climate Change Solutions Fund' set up at J.P. Morgan is another attempt to facilitate climate action (J.P. Morgan, 2022). Innovative business models that are sustainable and provide economic benefits to smallholders who adopt CNRFS must be developed for different business settings. In this process, eco-innovation will become the key to enabling a sustainable transition and green growth in the future by encompassing both economic and environmental values (OECD, 2012). Currently, the pace of these innovations is not adequate to drive the desired changes.

Assessing carbon storage and GHG emission reductions

Soil carbon sequestration and GHG emission reductions from the implementation of CNRFS will depend on several factors, including soil type, management capabilities of farmers and incentives that follow. Quantification of GHG emissions and soil carbon will be required to provide the benefits to farmers that are responsible for mitigation efforts. Since agriculture is a nonpoint source with dispersed biological emissions, criteria used in the energy markets to estimate carbon credits don't fit with agriculture. Standard and measurable indicators are to be developed for agriculture and forestry sectors and supported by easy-to-use robust models that will make it easier for monitoring and measurement. Harmonized and transparent methodologies, including remote monitoring systems for measurement and reporting of GHGs and sinks in different farming systems, as well as tools to support decision-making at farm and landscape level will be necessary (EIP-AGRI, 2021).

There are already some scientific models that provide values for different soil and climatic conditions, but more research and development must be taken up to develop new tools that can be used on small farms (WUR, 2019). The values should also reflect the time scale for storage based on which credits can be estimated. A built-in system to issue carbon credits and payments must be incorporated so that big and small farmers are part of the carbon economy in the future. Such an integrated system to measure soil carbon, GHG emission reductions and estimate the corresponding total or net carbon credits will be challenging but necessary for the agriculture and forestry sectors due to the multiple factors that influence the emissions in these sectors and the trade-offs that must be taken into consideration.

What can motivate farmers?

Without economic gains farmers will not be motivated to implement CNRFS. Better price for their products, incentives for implementing CNRFS that may lead to additional costs in production, insurance and risk coverage where needed, ready access to seeds and other inputs, access to technology and capacity building would be necessary. It is important to communicate potential co-benefits (rather than opportunities to earn compensation or carbon credits) and long-term benefits to increase farmers' engagement in carbon sequestration activities (Dumbrella et al., 2015). Farmers will face multiple challenges while changing the farming practices, for example, to change tillage, adopt new crop rotations, manure- and residue-management practices; and at the same time deal with risks from climate and extreme weather and unstable markets for their products (WRI, 2019). The dispersed nature and size of smallholders' farms will make it even more challenging for practical implementation of CNRFS and measurements of emission reductions unless there are easy-to-use monitoring and reporting mechanisms and adequate incentives that might accelerate the transition to carbon-neutral production (Lowder et al., 2016). In his study, Gullickson (2021) reported that diverse views prevail among farmers regarding climate change and changing farming practices to reduce GHG emissions. However, carbon credits and integrating farmers into carbon markets may help to bridge diverse views and motivate farmers.

Conclusions and way forward

The next leap we take to improve agricultural productivity must be carefully planned keeping sustainability as the central focus, combining nature-based and technology-based solutions, addressing both climate adaptation and mitigation simultaneously. Reviving some of the simple, low-cost and old agricultural practices based on indigenous knowledge systems together with the newly introduced technologies must pave the way forward to build a CNRFS structure suitable for different farm sizes, big and small, and different agroecological settings. It must be accompanied by a systemic upscaling approach, farmer organization, gender integration and collective action at different levels.

Investments to support the climate action must be effectively used, pooling resources, bringing different agencies (agriculture, forestry, energy) and sectors (public and private) together to cooperate, creating common infrastructure and work under one umbrella wherever possible. As discussed earlier, countries must see climate crisis as an opportunity for innovation action and address the challenges through development programs that could be 'climate proofed' rather than waiting for new programs and investments. In this way, the environmental and climate priorities no longer need to be put secondary on the development agenda. Every country should prepare a clear climate strategy and well-developed evidence-based NDCs highlighting climate actions, and detailed plans for implementing them.

Furthermore, innovative research and development initiatives must be promoted by governments and international agencies to explore new solutions and at the same time provide common platforms for sharing the knowledge developed and collectively addressing the climate crisis. In this way, duplication of efforts to develop research and knowledge can be avoided so that the limited funding is efficiently used by countries and all actors in the climate action.

Farmers, especially smallholders in developing countries, must be given access to necessary inputs, training and knowledge, and included in the carbon economy, thereby incentivizing their efforts to reduce GHGs. Countries need to increase their transparency in the implementation of CNRFS, measuring reporting and verification systems for emissions from agriculture and forestry sectors that are difficult to monitor. As most of us agree, our actions today both at individual and collective levels in addressing the climate crisis will determine the future of the humanity and the planet.

References

Asian Development Bank (ADB) (2021) 'Transforming agriculture in Asia'. Available at: https://www.adb.org/sites/default/files/publication/726556/ado2021-update-theme-chapter.pdf
Barth, B. (2016) 'Carbon farming: Hope for a Hot Planet'. Available at: https://modernfarmer.com/2016/03/carbon-farming/
Becker, S., Chameeva, T.B. and Jaegler, A. (2020) 'The carbon neutrality principle: A case study in the French spirits sector'. Available at: https://www.sciencedirect.com/science/article/pii/S0959652620327864

CGIAR (2020) 'Zero tillage to reduce air pollution in India'. Available at: https://www.cgiar.org/innovations/zero-tillage-to-reduce-air-pollution-in-india/

Dainese, M., Martin, E.A., Aizen, M.A., Albrecht, M., Bartomeus, I., Bommarco, R. and Steffan-Dewenter, I. (2019) 'A global synthesis reveals biodiversity-mediated benefits for crop production', *Science Advances*, vol. 5(10), pp. 121.

Delbeke, J. and Vis, P. (2016) 'EU Climate Policy Explained'. Available at: https://ec.europa.eu/clima/system/files/2017-02/eu_climate_policy_explained_en.pdf

Dumbrella, N.P., Kragtabc, M.E. and Gibsonab, F.L. (2015) 'What carbon farming activities are farmers likely to adopt? A best-worst scaling survey'. Available at: https://api.research-repository.uwa.edu.au/ws/portalfiles/portal/35085559/AAM_What_carbon_farming_activities.pdf

EIP-AGRI (2021) 'EIP-AGRI Workshop Towards carbon neutral agriculture'. Available at: https://ec.europa.eu/eip/agriculture/sites/default/files/eip-agri_ws_carbon_neutral_agriculture_final_report_2021_en_lr.pdf

European Commission (2021) 'Horizon Europe'. Available at: https://ec.europa.eu/info/research-and-innovation/funding/funding-opportunities/funding-programmes-and-open-calls/horizon-europe_en

EEA (European Environment Agency) (2021) 'Nature-based solutions in Europe: Policy, knowledge and practice for climate change adaptation and disaster risk reduction'. Available at: https://publications.europa.eu/en/publications

Fazekas, A., Puig, J. and Salinas, A.G. (2021) 'How are NDC INVEST and the CIF supporting countries to reach their climate goals?'. Available at: https://blogs.iadb.org/sostenibilidad/en/how-are-ndc-invest-and-the-cif-supporting-countries-to-reach-their-climate-goals/

Gan, Y., Liang, C., Chai, Q., Lemke, R.L., Campbell, C.A. and Zenter, R.P. (2014) 'Improving farming practices reduces the carbon footprint of spring wheat production'. Available at: https://www.nature.com/articles/ncomms6012

Government of India (2015) 'Soil Health Card'. Available at: https://www.soilhealth.dac.gov.in/

Gullickson, G. (2021) 'How carbon may become another crop for farmers'. Available at: https://www.agriculture.com/farm-management/programs-and-policies/how-carbon-may-become-another-crop-for-farmers

Howarth, C. and Painter, J. (2016) 'Exploring the science–policy interface on climate change: The role of the IPCC in informing local decision-making in the UK', *Palgrave Communications*, vol. 2, pp. 16058.

IFAD (2021) 'Examining the climate finance gap for small-scale agriculture'. Available at: https://www.ifad.org/documents/38714170/42157470/climate-finance-gap_smallscale_agr.pdf/34b2e25b-7572-b31d-6d0c-d5ea5ea8f96f

IGI (2021) 'IGI Launches New Research in Net-Zero Farming and Carbon Capture'. Available at: https://innovativegenomics.org/news/net-zero-farming-carbon-capture/

IISD (2019a) 'Why gender matters in climate change adaptation'. Available at: https://www.iisd.org/articles/gender-climate-change

IISD (2019b) 'Strengthening science-policy interface can help meet climate, sustainable development goals'. Available at: https://sdg.iisd.org/news/strengthening-science-policy-interface-can-help-meet-climate-sustainable-development-goals/

International Science Council (2021) 'Four considerations for accelerating progress on climate change at the science-policy interface'. Available at: https://council.science/current/blog/accelerating-progress-on-climate-change-at-the-science-policy-interface/

INRAE (2021) 'For more climate neutral and climate-resilient farms across Europe'. Available at: https://www.inrae.fr/en/news/more-climate-neutral-and-climate-resilient-farms-across-europe

IPCC (2021) 'Climate change widespread, rapid and intensifying'. Available at: https://www.ipcc.ch/2021/08/09/ar6-wg1-20210809-pr/

IPCC (2022) 'Climate change 2022: Impacts, adaptation and vulnerability'. Available at: https://report.ipcc.ch/ar6wg2/pdf/IPCC_AR6_WGII_FinalDraft_FullReport.pdf

JPI Climate (2021) 'Integrating science into climate change policy making: An overview of inspiring real-world practices'. Available at: http://www.jpi-climate.eu/media/default.aspx/emma/org/10903832/SINCERE+WP3+-+Best+Practice+Science+Policy+def.pdf

Kingwell, R. (2021) 'Making agriculture carbon neutral amid a changing climate: the case of South-Western Australia'. *Land*, vol. 10, pp. 1259.

Lehner, P.H. and Rosenberg, N.A. (2021) 'Farming for our future: The science, law, and policy of climate-neutral agriculture'. Available at: https://www.eli.org/eli-press-books/farming-our-future-science-law-and-policy-climate-neutral-agriculture

Lowder, S. K., Skoet, J. and Raney, T. (2016) 'The number, size, and distribution of farms, smallholder farms, and family farms worldwide', *World Development*, vol. 87, pp.16–29.

Morgan, J.P. (2022) 'Climate change solutions fund'. Available at: https://am.jp-morgan.com/no/en/asset-management/adv/investment-themes/sustainable-investing/capabilities/climate-change-solutions-fund

Nagothu, U.S. (2015) The future of food Security: summary and recommendations. In: Nagothu, U.S. (ed.) *Food Security and Development: country case studies*. London: Routledge, pp. 274.

Nagothu, U.S., Bloem, E. and Andrew B. (2018) 'Agricultural development and sustainable intensification: technology and policy innovations', In: Nagothu, U.S (ed.) *Agricultural Development and Sustainable Intensification, Technology and Policy Challenges in the Face of Climate Change*, London: Routledge, pp. 1–22.

Nkiaka, E. and Lovett, J.C. (2019) 'Strengthening the science-policy interface for climate adaptation: stakeholder perceptions in Cameroon', *Regional Environmental Change*, vol. 19: pp. 1047–1057.

OECD (2009) 'Integrating climate change adaptation into development cooperation'. Available at: https://www.oecd.org/env/cc/44887764.pdf

OECD (2012) 'The future of eco-innovation: the role of business models in green transformation'. Available at: https://www.oecd.org/innovation/inno/49537036.pdf

OECD (2019) 'Enhancing climate change mitigation through agriculture'. Available at: https://www.oecd-ilibrary.org/sites/16af156c-en/index.html?itemId=/content/component/16af156c-en

OECD (2021a) 'International programme for action on climate change'. Available at: https://www.oecd.org/climate-action/ipac/the-annual-climate-action-monitor-5bcb405c/

OECD, (2021b) 'Enhancing climate change mitigation through agriculture'. Available at: https://www.oecd.org/publications/enhancing-the-mitigation-of-climate-change-though-agriculture-e9a79226-en.htm

Reardon, T., Echeverria, R., Berdegué, J., Minten, B., Liverpool-Tasie, S., Tschirley, D. and Zilberman, D. (2019) 'Rapid transformation of food systems in developing regions: Highlighting the role of agricultural research & innovations', *Agricultural Systems*, vol. 172, pp. 47–59.

Rural Hub (2021) 'Practical ways to cut greenhouse gas emissions from farms'. Available at: https://rural.struttandparker.com/article/practical-ways-to-cut-greenhouse-gas-emissions-from-farms/

SIANI (2020a) 'Seven way to make crop production climate neutral'. Available at: https://www.siani.se/news-story/seven-ways-to-make-crop-production-climate-neutral/

SIANI (2020b) 'Nature based solutions – tools to manage the climate crisis'. Available at: https://www.siani.se/news-story/nature-based-solutions-parts-of-the-solution-to-the-climate-crisis/

SIPRI (2019) 'Climate crisis in times of geopolitical crises: what ways forward?'. Available at: https://www.sipri.org/commentary/blog/2019/climate-security-times-geopolitical-crises-what-ways-forward

The Commonwealth (2021) 'Gender integration for climate action: a review of commonwealth member country nationally determined contributions'. Available at: https://thecommonwealth.org/sites/default/files/inline/Gender.pdf

UNCTAD (2021) 'Carbon emissions anywhere threaten development everywhere'. Available at: https://unctad.org/news/carbon-emissions-anywhere-threaten-development-everywhere

UNEP (2021) 'Climate Action'. Available at: https://www.unep.org/news-and-stories/story/cop26-ends-agreement-falls-short-climate-action

UNFCCC (2017) 'China and India lead global energy transition'. Available at: https://unfccc.int/news/china-and-india-lead-global-renewable-energy-transition

UNFCCC (2019a) 'Climate action and support trends'. Available at: https://unfccc.int/sites/default/files/resource/Climate_Action_Support_Trends_2019.pdf

UNFCCC (2019b) 'NDC Spotlight'. Available at: https://unfccc.int/process/the-paris-agreement/nationally-determined-contributions/ndc-spotlight

United Nations (2020) 'Sustainable Development Goals'. Available at: https://www.un.org/sustainabledevelopment/climate-change/

UNDP (2021) 'NDC Support Program'. Available at: https://www.ndcs.undp.org/content/-ndc-support-programme/en/home.html

United Nations (2021) 'COP 26 A snapshot of the agreement'. Available at: https://unric.org/en/cop26-a-snapshot-of-the-agreement/

Van der Pol, L., Manning, D., Cotrufo, F. and Machmuller, M. (2021) 'To make agriculture more climate friendly, carbon farming needs clear rules'. Available at: https://theconversation.com/to-make-agriculture-more-climate-friendly-carbon-farming-needs-clear-rules-160243

World Bank, (2021) 'NDC Support Facility'. Available at: https://www.worldbank.org/en/programs/ndc-support-facility

WRI (2019) 'Reduce Greenhouse gas emissions from Agriculture'. Available at: https://research.wri.org/wrr-food/course/reduce-greenhouse-gas-emissions-agricultural-production-synthesis

WRI (2021) 'National climate action under the Paris agreement'. Available at: https://www.wri.org/ndcs

WUR (2019) 'Farmers working on climate neutral agriculture'. Available at: https://www.wur.nl/en/show/farmers-working-on-climateneutral-agriculture.htm

Acknowledgements

I would like to thank all the contributors of the various book chapters, particularly the lead authors who have spent their time in coordinating and drafting the chapters. This book would not have been possible without the support from the Norwegian Institute of Bioeconomy Research, Ås, Norway.

The support from my wife Shanthi, daughter Shreya and my son Sankalp during the drafting and compilation process helped me to complete the book on time, despite the challenges we faced at home during this period. Finally, I want to acknowledge the inspiration I have got from my late father who passed away in November 2020 when I just started to prepare the book manuscript.

Index

Note: **Bold** page numbers refer to tables; *italic* page numbers refer to figures and page numbers followed by "n" denote endnotes.

For Product Safety Concerns and Information please contact our EU
representative GPSR@taylorandfrancis.com
Taylor & Francis Verlag GmbH, Kaufingerstraße 24, 80331 München, Germany

www.ingramcontent.com/pod-product-compliance
Lightning Source LLC
Chambersburg PA
CBHW060253220326
41598CB00027B/4080